Progress ... .-. -. ....-

MW00724052

Project Management at the Department of Energy

2001 Assessment

Committee for Oversight and Assessment of
U.S. Department of Energy Project Management

Board on Infrastructure and the Constructed Environment

Division on Engineering and Physical Sciences

National Research Council

NATIONAL ACADEMY PRESS
Washington, D.C.

NATIONAL ACADEMY PRESS • **2101 Constitution Avenue, N.W.** • **Washington, DC 20418**

NOTICE: The project that is the subject of this report was approved by the Governing Board of the National Research Council, whose members are drawn from the councils of the National Academy of Sciences, the National Academy of Engineering, and the Institute of Medicine. The members of the committee responsible for the report were chosen for their special competences and with regard for appropriate balance.

This study was funded by the U.S. Department of Energy, Contract Number DE-AM01-99PO8006. All opinions, findings, conclusions, and recommendations expressed herein are those of the authors and do not necessarily reflect the views of the Department of Energy.

International Standard Book Number 0-309-08280-3

Additional copies of this report are available for sale from National Academy Press, 2101 Constitution Avenue, N.W., Lockbox 285, Washington, DC 20055; (800) 624-6242 or (202) 334-3313 (in the Washington metropolitan area); also available online at <http://www.nap.edu>.

THE NATIONAL ACADEMIES

National Academy of Sciences
National Academy of Engineering
Institute of Medicine
National Research Council

The **National Academy of Sciences** is a private, nonprofit, self-perpetuating society of distinguished scholars engaged in scientific and engineering research, dedicated to the furtherance of science and technology and to their use for the general welfare. Upon the authority of the charter granted to it by the Congress in 1863, the Academy has a mandate that requires it to advise the federal government on scientific and technical matters. Dr. Bruce M. Alberts is president of the National Academy of Sciences.

The **National Academy of Engineering** was established in 1964, under the charter of the National Academy of Sciences, as a parallel organization of outstanding engineers. It is autonomous in its administration and in the selection of its members, sharing with the National Academy of Sciences the responsibility for advising the federal government. The National Academy of Engineering also sponsors engineering programs aimed at meeting national needs, encourages education and research, and recognizes the superior achievements of engineers. Dr. Wm. A. Wulf is president of the National Academy of Engineering.

The **Institute of Medicine** was established in 1970 by the National Academy of Sciences to secure the services of eminent members of appropriate professions in the examination of policy matters pertaining to the health of the public. The Institute acts under the responsibility given to the National Academy of Sciences by its congressional charter to be an adviser to the federal government and, upon its own initiative, to identify issues of medical care, research, and education. Dr. Kenneth I. Shine is president of the Institute of Medicine.

The **National Research Council** was organized by the National Academy of Sciences in 1916 to associate the broad community of science and technology with the Academy's purposes of furthering knowledge and advising the federal government. Functioning in accordance with general policies determined by the Academy, the Council has become the principal operating agency of both the National Academy of Sciences and the National Academy of Engineering in providing services to the government, the public, and the scientific and engineering communities. The Council is administered jointly by both Academies and the Institute of Medicine. Dr. Bruce M. Alberts and Dr. Wm. A. Wulf are chairman and vice chairman, respectively, of the National Research Council.

Acknowledgment of Reviewers

This report has been reviewed in draft form by individuals chosen for their diverse perspectives and technical expertise, in accordance with procedures approved by the NRC's Report Review Committee. The purpose of this independent review is to provide candid and critical comments that will assist the institution in making its published report as sound as possible and to ensure that the report meets institutional standards for objectivity, evidence, and responsiveness to the study charge. The review comments and draft manuscript remain confidential to protect the integrity of the deliberative process. We wish to thank the following individuals for their review of this report:

Philip R. Clark, Nuclear Corporation (retired),
Fletcher H. (Bud) Griffis, Polytechnic University,
Henry J. Hatch, U.S. Army Corps of Engineers (retired),
Elvin R. Heiberg III, Heiberg Associates, Inc.,
Bradley C. Moore, Ohio State University, and
Richard N. Zare, Stanford University.

Although the reviewers listed above have provided many constructive comments and suggestions, they were not asked to endorse the conclusions or recommendations, nor did they see the final draft of the report before its release. The review of this report was overseen by Charles B. Duke (NAE), Xerox Research and Technology. Appointed by the National Research Council, he was responsible for making certain that an independent examination of this report was carried out in accordance with institutional procedures and that all review comments were carefully considered. Responsibility for the final content of this report rests entirely with the authoring committee and the institution.

Contents

Executive Summary

The Department of Energy (DOE) is engaged in numerous multimillion- and even multibillion-dollar projects that are one of a kind or first of a kind and require cutting-edge technology. The projects represent the diverse nature of DOE's missions, which encompass energy systems, nuclear weapons stewardship, environmental restoration, and basic research. Few other government or private organizations are challenged by projects of a similar magnitude, diversity, and complexity. To complete these complex projects on schedule, on budget, and in scope, the DOE needs highly developed project management capabilities.

This report is an assessment of the status of project management in the Department of Energy as of mid-2001 and the progress DOE has made in this area since the National Research Council (NRC) report *Improving Project Management in the Department of Energy* (Phase II report) was published in June 1999 (NRC, 1999). The Phase II report findings and recommendations are reproduced as Appendix C. The findings presented in this report reiterate and expand on those given in the committee's January 2001 interim letter report, *Improved Project Management in the Department of Energy* (NRC, 2001), reproduced here as Appendix D.

The Phase II report estimated that DOE projects costed taxpayers 50 percent more than comparable projects would cost if performed by the private sector or other government agencies, in large part because DOE did not use industry-standard best practices for project management. The Phase II report recommended, inter alia, that DOE develop policies, procedures, models, tools, techniques, and standards; train staff in their use; and require their use on DOE projects. It recommended further that DOE should develop and deploy a comprehensive project management system with clear definition of the specific roles and responsibilities of all parties associated with a project.

As noted in the interim letter report, the department has taken a number of positive steps since the Phase II report. On June 25, 1999, subsequent to the release of that report, the deputy secretary, as the DOE chief operating officer, issued a memorandum announcing a project management reform initiative. This memorandum directed a number of actions to be taken to improve project management capability. These included the formation of the Office of Engineering and Construction Management (OECM) in the office of the chief financial officer (CFO) and the formation and strengthening of project management support offices (PMSOs) in the three major program secretarial offices (PSOs). On January 3, 2000, the deputy secretary issued an interim instruction to serve as policy guidance on critical decisions by acquisition executives (AEs) and the Energy Systems Acquisition Advisory Board (ESAAB) and on the conduct of corporate-level performance reviews. On June 10, 2000, DOE issued Policy P413.1, which addresses project management accountability, the establishment of project management organizations, project management tools, and training of personnel. DOE Order O413.3 was issued October 13, 2000, to implement the DOE policy document. O413.3 covered department policies on project management; provision for project engineering and design (PED) funding for preconstruction planning; reestablishment of the ESAABs; and other matters related to the management and oversight of DOE projects. Finally, the *Program and Project Management* (PPM) manual and a companion volume, *Project Management Practices* (PMP), were released in draft form in October 2000.

The body of this report addresses certain specific areas that the committee believes needed greater definition and follow-up since the Phase II report. Not all the findings and recommendations in the Phase II report are repeated in the body of this report, although the committee continues to endorse them, so this report should be used in conjunction with the Phase II report.

OVERARCHING ISSUES

The new secretary of energy, the deputy secretary, and the CFO were briefed on the recommendations of the Phase II report (NRC, 1999) and the 2001 letter report (NRC, 2001). All of these executives expressed their intention to gain control over DOE projects and to change the prevalent project management culture within DOE and its contractors through active management attention and oversight. Although DOE senior management has expressed the intent to achieve early results, it is too soon to see any effects from these management changes, as the new deputy secretary was confirmed only in June 2001. Accordingly, any policies and practices put in place by the new management team at DOE lie outside the scope of the present report, and their assessment must be deferred to a later time. Nevertheless, the committee is optimistic that if these stated intentions are carried through, there can be positive change at DOE.

The committee believes the efforts of the deputy secretary, the CFO, OECM, and the PMSOs have unquestionably raised awareness of the importance of good project management within the department. Briefings by DOE officials indicate that new steps are being undertaken at all levels of the department. This is encouraging, and if the efforts continue, they can form the foundation of a coherent project management approach for the department. By and large, these initiatives are considered by the committee to be steps toward improving project performance. However, change has been inordinately slow, and the committee has found no evidence that DOE project management practice and performance in the field have actually improved.

The committee believes that the PMSOs in the Environmental Management (EM), National Nuclear Security Agency (NNSA), and Office of Science (SC) PSOs are having a positive influence on project management in their programs. The committee recognizes the need for tailoring project management programs to the different PSO mission requirements. However, some PSOs are improving faster than others, and progress could be accelerated by stronger central oversight and support using OECM as a unifying organization to assist the PMSOs and the PSO projects and to validate the results.

Department-wide, recognition of project management deficiencies and of the effectiveness of initiatives for improvement has been inconsistent. The PMSO in NNSA has been particularly active since formation of the agency. Excessive administrative impediments and time required to get the OECM up and running, along with the inadequate resources assigned to it, are symptoms of cultural resistance to change and the lack of a sense of urgency. The committee is aware that not all of the recommendations in the Phase II report could have been accomplished in the 2 years since it was released, but it does believe that much more progress could have been made. The committee recommends that OECM should be budgeted, staffed, and empowered to become the center of excellence in project management and the central manager for oversight and approval of all capital projects in DOE.

DOE continues to rely heavily on contractors, not only for the management of projects, which is to be expected, but also for project justification and definition of scope. DOE does not directly manage projects, unless the contractor is simply working on a time and materials basis, as do some management and operations (M&O) contractors. Although many projects are managed successfully, especially those identified as outstanding examples in the OECM October 2000 project management workshop awards, DOE is not in control of many of its projects and in some cases has virtually abdicated its role of owner in project oversight and management of contracts and contractors. As stated in the Phase II report, DOE needs to become proficient in the role of owner; and DOE project managers should become knowledgeable owner's representatives. The committee has seen little evidence that DOE is aware of the distinction between the owner's and contractor's role, let alone acting upon it.

Two of the overarching issues that particularly concern the committee are the lack of strategic planning and the lack of a system for process improvement. The first step toward process definition has been taken in O413.3, but much remains to be done. A consistent, documented process and procedures for efficient project planning, justification, and execution are essential. Treatment of each project on a unique, ad hoc basis continues and should be seen as unacceptable. The committee believes that each PSO should develop its own strategic plan and that project justification as required for approval of mission need (CD-0) should be based on an assessment of the congruity of project descriptions with these strategic plans. The committee recommends that DOE initiate a program for improving the project management process, following an established statistical process control procedure such as the well-known six-sigma process.

FRONT-END PLANNING

The committee has received no evidence that new projects are getting off to a better start than before the Phase II report was issued. The establishment of PED funding and the development of the performance baseline at 25 to 30 percent design completion are positive steps, but the committee's review of new projects authorized in FY2000 and FY2001 uncovered little documentation of mission need, strategic requirements, scope justification, risk assessment, and the other early steps that are necessary for project success. Industry studies consistently correlate project performance with the quality of front-end planning (CII, 2000). The committee reiterates the Phase II finding that improvement in project performance will require improvement in the front-end planning process, and that any improvement in front-end planning will require positive, aggressive action by all responsible parties. It should be noted that all the projects reviewed by or presented to the committee were initiated prior to the activation of the OECM, the establishment of the PMSOs, and the issuance of O413.3. Therefore, this negative assessment does not mean that these initiatives are ineffective. What it does indicate, based on the evidence provided to the committee, is that the DOE process has not noticeably improved since the 1999 Phase II report and probably will not improve until the reforms have been implemented and have become effective.

The committee recommends that OECM assure that all program offices have a documented front-end planning process that meets the intent of O413.3, and that the information be used as input for Energy Systems Acquisition Advisory Boards (ESAABs) and ESAAB-equivalent readiness reviews. The outcomes of these reviews should be documented and used to assess project performance and progress in improving project planning. Without immediate improvement in the planning knowledge and skills of personnel and more management emphasis on improving the planning process, projects will continue to have inadequate front-end planning. To overcome the lack of internal skilled project planners and the delays in training, and to bridge the gap until a training program takes effect, the committee recommends that DOE should engage a cadre of experienced project planners with a wide variety of planning capabilities and prior experience in the different project types, including high-risk projects. These individuals should be a part of the initial integrated project teams and should assist the project originators (as internal consultants) in getting front-end planning done correctly, even before CD-0.

RISK MANAGEMENT

Risk management is probably the most difficult aspect of project management, and for many DOE projects it is also the most critical. Discussions with DOE project managers have revealed that DOE has not yet established effective risk-management methodologies or systems, and managers lack the tools and training to adequately address risk management for projects with high levels of uncertainty. The acquisition risk management (ARM) pilot program by the DOE Contract Reform Office is beginning to address many risk management issues and has defined three iterative phases of risk management for DOE projects:

1. Risk identification,
2. Risk analysis and evaluation, and
3. Risk response.

The committee believes that all three phases of risk management are critical to effective project management at DOE. The committee also believes that DOE should undertake a department-wide assessment of risks for ongoing projects and adopt procedures for program-wide management of contingencies and management reserves.

The committee has found that there is no consistent system for evaluating the relative risks of projects with respect to scope, cost, or duration, which means that the deputy secretary, the CFO, and the PSO managers have no objective basis for knowing which projects are riskier (and therefore require more management attention) than others. Probability analysis has been used and misused as a tool for risk analysis. However, the failure to accurately identify the root causes of risk, including technical, environmental, and human factors, and the potential for common mode failure has led to underestimating risks on many DOE projects. The committee recommends that DOE should identify or develop personnel with the ability to perform qualitative and quantitative risk assessments and assign them to work with personnel with in-depth understanding of a given project to undertake risk management. Independent assessors or reviewers who are not project proponents should separately evaluate internal project risk assessments for reasonableness of assumptions, estimates, and results. Risk mitigation and management plans should be prepared to deal with any significant risks identified, especially risks due to common modes or root causes.

The committee recommends that DOE should develop more expertise and improved tools for risk management. Nontraditional and innovative approaches, tools, and methods should be investigated for adaptation to DOE project conditions and use in DOE risk management, including those cited in this report and in the Phase II report (NRC, 1999, Appendix B), such as systems analysis, event trees, causal loop diagrams, system dynamics, stochastic simulation, and other approaches that have been tested and shown to be valuable on similar projects or in addressing similar challenges.

PROJECT REPORTING AND OVERSIGHT

Effective oversight of project performance is dependent on systematic and realistic reporting of project performance data. Each PSO has its own active project reporting systems, and OECM has completed a specification and beta testing for a department-wide project analysis and reporting system (PARS). The committee believes that PARS should be designed so that it supports the data analysis needed by project managers to evaluate their projects' performance as well as the oversight needs of the PSOs, OECM, the CFO, and the deputy secretary. The database should also provide the information needed for process improvement through statistical process control and benchmarking future projects.

An effective earned value management system (EVMS) is needed as the source of data for PARS and data analysis for project management and project oversight. To create an effective department-wide earned value management and reporting system, a consistent accrual accounting system is needed for all projects.

The committee believes that DOE management needs to be able to detect potentially adverse trends in project progress and distinguish them from mere random fluctuations in progress reporting. EVMS provides data that can be used to gain some very valuable insights into the health of a project and can predict the probable outcome. EVMS data can also allow useful insights into the conduct of the work, particularly when it is reported and analyzed to evaluate period-to-period trends. The committee recommends that DOE utilize EVMS data to calculate the incremental and cumulative cost performance index, schedule performance index, and contingency utilization index for each reporting period and that it produce process control charts to analyze and improve project performance.

INDEPENDENT REVIEWS

The committee finds that DOE has made substantial progress in the implementation of reviews and resultant corrective action plans and in the formalization and institutionalization of the review process. It continues to recommend the use of formalized assessments of management, scope, cost, and schedule at appropriate stages—from determining need for the project to determining readiness for construction—as well as regular performance reviews.

The committee examined the internal review processes documented by the three PMSOs and noted some inconsistency in general approach and the degree of recent process improvement. It strongly endorses a department-wide manual to achieve consistency in process, nomenclature, and reporting. Department-wide procedures would allow the lessons learned in reviews to gain broader recognition and would facilitate the ability of reviewers to participate in reviews in multiple organizations. Any differences or distinct features pertinent to a particular program could be identified and articulated in the text or in appendixes.

The committee is concerned that mandatory review for projects between $5 million and $20 million total project cost may be consuming too many resources and diverting too much management attention relative to the value it adds. The committee understands that although O413.3 refers to the $5 million level, OECM intends to waive the requirement if it believes a review would not be cost effective.

ACQUISITION AND CONTRACTING

The committee has reaffirmed the integral role that acquisition planning and acquisition and contracting techniques play in successful project management, as noted in both of its previous reports (NRC 1998, 1999). This role is particularly critical in the DOE environment, where some 90 percent of the department's budget is expended by contract (GAO, 2001). The committee stresses the importance of developing and employing contracting methods, such as performance-based contracting (PBC), that ensure accountability, adequately address risk, and focus the government and the contractor on achieving the outcomes sought.

Successful PBC is based on defining existing conditions, specific requirements, and the desired results or outcomes, along with objective, meaningful, and measurable performance and quality standards. In addition, incentives are used to focus contractor efforts and to reward success. In a successful performance-based contract, expectations should be made clear, with agency and contractor teams working together in a business partnership to achieve well-defined and measurable results.

The integrated project team (IPT) concept included in DOE Order O413.3 is an essential element in implementing a performance-based approach. The committee strongly supports the use of IPTs and suggests a PBC methodology for forming teams and developing performance metrics and incentives. The committee recommends intensive PBS training for IPTs and recommends that OECM should, in the near term, bring on board a cadre of experts skilled in performance-based contracting to provide technical assistance to IPTs responsible for new major system initiatives.

The committee recommends that tailoring contracting approaches to use fixed-price and performance-based methods where practicable will assist the department in getting the most cost-effective results and will also result in greater competition. In addition, the department should continue to explore other innovative commercial contracting approaches to meet its needs.

DOCUMENTATION OF PROJECT MANAGEMENT POLICIES AND PROCEDURES

The OECM is to be commended on the significant progress in getting Policy P413.1 and Order O413.3 produced and published; however, the committee has noted shortcomings in these documents and in the draft documents *Program and Project Management* (PPM) manual and *Project Management Practices* (PMP). The committee observed that O413.3 has proven to be effective in defining and implementing a number of fundamental and beneficial changes for the department that will improve long-term project performance; however, there are several clarifications, improvements, and adjustments that, while not changing the basic policy, would improve it.

The draft PPM and PMP were overwhelming in their detail in some sections, which in some instances was superfluous and in others misleading; major issues, by contrast, received very little attention. Some important issues were missing entirely, such as team alignment and teamwork procedures for including stakeholder and public participation; project scope definition; and control of scope change. The committee recommends that the PPM and PMP texts should be tailored to specific DOE requirements. It should be made clear which parts of the text constitute DOE required procedures and which parts reflect general advice on good project management practices. OECM should assure that policies and required procedures add value by streamlining the process and improving project performance. Examples should be given, where possible, to illustrate the application of procedures and the necessary documentation. The examples should have adequate explanations and represent realistic project situations. Over time, a set of templates and case studies should be built up.

More important, and of much greater concern to the committee, is the pervasive slowness of change at DOE in response to these problems. There continues to be excessive dependence on written policies and procedures to effect change. Although O413.3 was useful for defining DOE policy, guidelines, directives, and orders from DOE headquarters, even if they are good, will not by themselves effect change. If paper guidelines had been adequate, there would have been no need for this committee or for its predecessor. In the final analysis, the effectiveness of any policies and procedures depends on a commitment by DOE leadership to the continuous improvement of DOE project performance by proven project management methods and techniques, as defined in departmental policies and procedures.

PROJECT MANAGER TRAINING AND DEVELOPMENT

In the more than 2 years since the 1999 NRC report recommended training for DOE project managers, no departmental training program is in place. Although an expensive study on human resources management is under way, no one has been trained under it, and the committee has been advised that training may not begin until 2003. The need for project management training by experienced project managers was apparent to DOE long before the Phase II report, and the lack of training is a major impediment to improved project management in the department.

In January 2001, the deputy secretary directed OECM to lead a 2-year effort to develop and implement the Project Management Career Development Program (PMCDP). To accomplish this goal, a task force was established that included representatives from PSO headquarters and field offices and experts from other federal agencies. The committee applauds the task force effort to create a program geared to developing the knowledge and skills needed by project managers to fulfill the missions of the agency. However, despite the fact that the final curriculum for project manager training will not be completed until next year, it is imperative that training not be neglected in the interim. The committee recommends that DOE should implement an immediate, accelerated training program to improve the knowledge, skills, and abilities of project managers to address recognized gaps while continuing the PMCDP planning effort. This should be accomplished by eliminating impediments and using current resources, as well as exploring creative and cost-effective nonclassroom alternatives. DOE management should budget the funds required to accomplish the projected training objectives and should persist in mandating the accomplishment of individual career development objectives.

CONCLUSIONS

As stated in the Phase II report (NRC, 1999) and the committee's later letter report (NRC, 2001), effective and accountable project management should be a priority for DOE and its leaders at all levels. Through actions taken to date, DOE has begun to address some of the core issues; however, a number of issues have not been resolved. The committee addresses these issues in detail in the findings and recommendations in this report. The committee will continue to look for long-term project management reforms, process improvement, and the resultant improvement in project performance.

REFERENCES

CII (Construction Industry Institute). 2000. Benchmarking and Metric Report. Austin, Tex.: Construction Industry Institute.

GAO (General Accounting Office). 2001. Major Management Challenges and Program Risks: Department of Energy. GAO/GAO-01-246. Washington, D.C.: Government Printing Office.

NRC (National Research Council). 1998. Assessing the Need for Independent Project Reviews in the Department of Energy. Washington, D.C.: National Academy Press.

NRC. 1999. Improving Project Management in the Department of Energy. Washington, D.C.: National Academy Press.

NRC. 2001. Improved Project Management in the Department of Energy. Letter report, January. Washington, D.C.: National Academy Press.

1

Introduction

BACKGROUND

Recurrent problems with U.S. Department of Energy (DOE) project performance in the 1990s raised questions on the part of congressional appropriations committees about the credibility of the practices and processes used by the department to procure and manage projects. In an effort to increase confidence in DOE's capital acquisition budget, the 105th Committee of Conference on Energy and Water Resources directed DOE to investigate establishing a project review process. DOE requested the assistance of the National Research Council (NRC), which resulted in the publication of the report *Assessing the Need for Independent Project Reviews in the Department of Energy*, also known as the Phase I report (NRC, 1998). That report found that poor project performance in the DOE was due, in part, to deficiencies in the department's procedures for initiating and managing projects.

Congress also directed DOE to undertake a review and assessment of its overall management structure and process for identifying, managing, designing, and constructing facilities (U.S. Congress, 1997). DOE again asked the NRC for assistance, this time to conduct an independent review and develop recommendations to improve DOE's management of projects. The NRC published its findings and recommendations as *Improving Project Management in the Department of Energy*, also known as the Phase II report (NRC, 1999). The Phase II report indicated that the problems in DOE project management were pervasive and ingrained in the culture of the department. It provided a set of findings and recommendations as a guide to improving project management and noted that the problems could not be resolved by any single change and that improvement would require a program of reform for the entire project management process.

DOE's diverse missions are supported by hundreds of projects resulting in annual expenditures of billions of dollars. Consequently, Congress has an ongoing concern about project management in the DOE and the need to assure American taxpayers that the nation's resources are effectively and efficiently managed.

SCOPE OF WORK

In response to a directive from the 106th Committee of Conference on Energy and Water Resources, DOE requested the NRC to appoint a committee to review and assess the progress made by the department in improving its project management practices, as recommended in the Phase II report, and conducting adequate external, independent project reviews. The principal goal of this effort is to review and comment on DOE's recent efforts to implement the recommendations in the Phase II report and improve its project management, including a review of the following:

- Specific changes implemented by the department to achieve improvement (e.g., organization, practices, training);
- An assessment of the progress made in achieving improvement; and
- The likelihood that improvement will be permanent.

This oversight and assessment is planned as a 3-year effort. It will include annual reports on DOE's accomplishments, the identification of problems needing additional attention, and recommendations for departmental actions.

The NRC appointed a committee under the auspices of the Board on Infrastructure and the Constructed Environment (BICE) to undertake the review and assessment of DOE project management. The committee is composed of 11 professionals with diverse experience in academic, government, and industrial settings and knowledge of project management and process improvement. Five members of the committee also participated in the Phase II review and assessment, and one member participated in both Phase I and Phase II. See Appendix A for biographies of the committee members.

The committee met five times from September 2000 to July 2001 to review and assess data on projects and project management procedures presented by the DOE project managers and representatives of the Office of Engineering and Construction Management (OECM), the project management support offices (PMSOs) in the Office of Environmental Management (EM), the National Nuclear Security Agency (NNSA), the Office of Science (SC), and the Albuquerque Operations Office (AO). Committee representatives also attended project management workshops and awards programs sponsored by OECM, EM, NNSA, and SC and met with DOE senior managers responsible for managing programs, establishing policies, and implementing project management reforms. The committee's findings and recommendations are based on briefings and documents provided by DOE. The committee's fact-finding efforts are listed in Appendix B.

ORGANIZATION OF THE REPORT

This is the committee's first annual report. It includes the committee's assessment of progress in improving project management in the DOE as of mid-2001 and provides additional discussions of issues the committee determined to be key factors affecting project management in the department. The body of this report addresses some of the issues raised in the Phase II report that the committee believes are most critical to improving project management. Not all the findings and recommendations in the Phase II report are repeated here, although the committee continues to endorse them, so this report should be used in conjunction with the Phase II report. The Phase II report findings and recommendations are reproduced as Appendix C. The findings and recommendations in this report reiterate and expand on those given in the committee's interim letter report, *Improved Project Management in the Department of Energy* (NRC, 2001), reproduced here as Appendix D.

The report is organized in nine chapters. The findings and recommendations in Chapters 2 through 4 are listed after each subtopic discussion. In Chapters 5 through 9 they are listed in a separate section at the end of the chapter. Chapter 1, "Introduction," includes background information on earlier project management oversight and assessment efforts conducted for DOE by the NRC and the scope of the current study. Chapter 2, "Overarching Issues," includes a discussion of the department's organizational structure, the role of senior management, the role of DOE managers as project owners, and department-wide policies and procedures for strategic planning and process improvement.

The succeeding chapters discuss in detail specific aspects of project management that the committee believes are most critical and DOE's efforts to improve project management. Chapter 3, "Front-End Planning," addresses an aspect of project management that correlates closely with project performance. It includes background information on the process and impact of front-end planning and assesses current front-end planning efforts in the DOE. There, the committee discusses possible actions that it believes will improve oversight, evaluation, and improvement of the front-end planning process.

Chapter 4, "Risk Management," addresses an issue identified by DOE project managers and the committee as one of the most critical and difficult components of project management. It assesses DOE efforts to improve risk management and provides technical background on methods the department could employ to identify, analyze, evaluate, and respond to the risks inherent in DOE projects. A department-wide approach to assessing the level of risk in ongoing projects and to managing risk in the department's portfolio of projects is also discussed.

Chapter 5, "Project Reporting and Oversight," assesses department-wide efforts in data collection and analysis. It discusses analysis techniques to evaluate project performance and the use of the data for benchmarking and process improvement. Additional information on the use of project performance data for statistical process control is provided as Appendix E.

Chapter 6, "Independent Reviews," assesses progress in planning, managing, and implementing external independent reviews (EIRs), independent cost estimates (ICEs), and internal project reviews (IPRs). It assesses the documentation of procedures, review-team qualifications, review requirements, and review evaluations for EM, SC, and NNSA .

Chapter 7, "Acquisition and Contracting," reviews progress in implementing the Phase II recommendation that the department should employ contracting methods that address risk and assign accountability. There, the committee emphasizes the importance of performance-based contracting and methods for achieving the department's contracting objectives.

Chapter 8, "Documentation of Project Management Policies and Procedures," assesses the effectiveness of DOE project management policies and procedures documents issued since publication of the Phase II report, in June 1999. It addresses general aspects of the content and organization of the documents and specific issues regarding value engineering, change management, and ISO 9000 that the committee considers in need of additional attention.

The report concludes with an assessment of DOE progress in establishing a department-wide training program for project managers, as well as criteria and standards for the selection and assignment of project managers. Chapter 9, "Project Manager Training and Development," assesses the Project Management Career Development Program currently being developed and the department's interim efforts to ensure that project managers have the knowledge, skills, and abilities needed to manage DOE projects. Alternative approaches to delivering training to DOE managers are also discussed.

REFERENCES

NRC (National Research Council). 1998. Assessing the Need for Independent Project Reviews in the Department of Energy. Washington, D.C.: National Academy Press.

NRC. 1999. Improving Project Management in the Department of Energy. Washington, D.C.: National Academy Press.

NRC. 2001. Improved Project Management in the Department of Energy. Letter report, January. Washington, D.C.: National Academy Press.

U.S. Congress. 1997. Committee of Conference on Energy and Water Development. HR 105-271. Washington, D.C.: Government Printing Office.

2

Overarching Issues

INTRODUCTION

The Phase II report (NRC, 1999) identified a number of problems in DOE project management and provided a set of recommendations for improving project management within the department. It noted that the problems were long-standing and pervasive and that there were no quick fixes that could make the department's project management as good as that of other agencies and private industry. The recommendations included a structure to support project managers and provide consistent methods and systems to drive a change in the culture at DOE in the direction of improved project performance. The report concluded that improvement in project management requires the full and continuing support of the secretary of energy and the deputy secretary to ensure that reforms are enacted throughout DOE.

The committee reported in January 2001 that DOE had undertaken a number of initiatives to improve project management and address some of the core issues; it also identified issues that had not been addressed (NRC, 2001). Since January 2001, the committee has continued to evaluate DOE project management policies and procedures and assess project performance. The committee believes that DOE project management will improve only if senior management, indeed managers at all levels, demonstrate their support and commitment by taking an active role in project reviews, accepting responsibility for management procedures and project performance, and clearly defining management expectations. This involvement should not be limited to projects that are in trouble but should begin in early project planning and be maintained throughout the life of the project.

The committee has continued to review project management procedures and performance to assess the potential for long-term, lasting improvement. After issuing its interim report in January 2001, the committee reviewed the performance of many specific projects in the Albuquerque Operations Office (AO), the documentation of new projects recently funded by Congress, and department-wide procedures for front-end planning, risk management, acquisition and contracting, project reviews, reporting project performance, documentation of policies and procedures, and personnel training and development. These issues are addressed in detail in the succeeding chapters. In the remainder of this chapter, the committee summarizes the overarching issues and provides recommendations for dealing with them.

INVOLVEMENT OF SENIOR MANAGEMENT

The secretary of energy, the deputy secretary, the CFO, and other senior members of the new DOE management team have taken the time to discuss their views on project management in DOE with representatives of the committee. The committee appreciates their time and interest in keeping the committee informed about the new management philosophy and developments at DOE. Based on these statements, the committee is cautiously optimistic that major improvements may be forthcoming in DOE project management. The committee's optimism is based on its understanding that the following will be components of the new program and project management approach in DOE:

- Increased project management discipline,
- Implementation of professional development for project managers,
- Greater emphasis on project justification and mission need,
- Definition of options and decision points for project termination or change,
- Greater emphasis on accountability and responsibility for project performance,
- Expanded roles and responsibilities for OECM in approving projects,
- Formal quarterly program reviews by the deputy secretary,
- Clearly defined expectations for project performance,
- Recognition of DOE's role as an owner,
- Formal, clearly defined project performance and management metrics,
- Change in DOE culture driven from the top,
- Attention by the secretary, deputy secretary, and CFO, and
- Consolidation of management and administration, contracting, and project management under the CFO.

Similar DOE management goals were endorsed by the committee in the Phase II report and the January 2001 letter report. The committee looks forward with great anticipation to the proposed changes in DOE project management and will follow up in future assessment reports. However, it reiterates its conviction that there is no quick fix for DOE project management problems; that all of the above steps should be executed; that many DOE projects are inherently difficult owing to their high uncertainty and complexity; and that any effort to improve DOE project management must effectively address the issues of procurement, acquisition, contracting, and management of contracts and contractors.

STRATEGIC PLANNING

The Phase II report was concerned not only with DOE doing projects right but also with doing the right projects. However, the committee was unable to find documented justification for most projects, including new projects, and so could only note the absence of identifiable strategic plans for DOE as a whole and for the PSOs individually. The *Paths to Closure* document by EM is in some ways a very long-range plan and perhaps comes closest to meeting this need (DOE, 1998). However, what is needed in addition are rolling 5-year plans, with budget projections, for all program secretarial offices (PSOs), which can be used to identify upcoming projects, why they are needed to fulfill mission goals and requirements, and how they fit in the DOE strategic objectives. In this way, strategic plans can form the basis for, and bridge into, conceptual plans for specific projects. It is impossible in most cases to determine if DOE projects comply with the Government Performance and Results Act of 1993 (GPRA) because of the department's unclear objectives and insufficient performance measures (GAO, 2001a). Without articulated mission needs and objectives, there is no basis for identifying and evaluating benefits. The committee believes that each PSO should develop its own strategic plan and that project justification for critical decision 0, approval of mission need (CD-0), should assess the congruity of project descriptions with these strategic plans. Strategic 5-year plans should be updated annually to assure that projects are aligned with the evolving missions of the department.

Finding. There are no PSO strategic plans defining long-range goals and objectives or mission needs, and documentation of project justification is almost entirely lacking or inadequate, so that it is impossible to assess whether the right projects are being done. Cost-benefit analysis and performance measurements required by GPRA cannot be performed effectively without effective strategic plans.

Recommendation. The PSOs should develop budget-based rolling 5-year strategic plans that identify the mission goals and objectives of the program, the projects necessary to achieve them, and the benefits to be expected from these projects.

OWNER'S ROLE

DOE continues to rely heavily on contractors not only for the management of projects, which is to be expected, but also for project justification and definition of scope. DOE does not directly manage projects unless the contractor is simply working on a time and materials basis, as do some management and operations (M&O) contractors. However, in some cases DOE has virtually abdicated its role of owner in the oversight and management of contracts and contractors. Although many projects are managed successfully, especially those identified as outstanding examples in the OECM October 2000 project management workshop awards, there are still large projects where DOE is not executing the role of owner with respect to oversight and management of contracts and contractors and is not in effective control of these projects. As stated in the Phase II report, DOE needs to become proficient in the role of owner; and DOE project managers should become knowledgeable owner's representatives. The committee has seen little evidence that DOE is aware of the distinction between the project management roles of owners and contractors, let alone of acting on it.

The committee notes that there has been independent confirmation of these findings in the form of the June 2001 report *Department of Energy: Follow-up Review of the National Ignition Facility* (GAO, 2001b). In the nearly 2 years since the public revelation of massive cost and schedule overruns, according to the GAO, this project still lacks a defined mission need and goals acceptable to the three national laboratories—Lawrence Livermore (LLNL), Sandia (SNL), and Los Alamos (LANL). Moreover, it does not have a fully staffed oversight office in DOE, does not have a technical risk assessment capability, relies on optimistic assumptions about operational issues, does not have essential predecessor activities under project control, and does not have an independent external review process. These are all elementary, basic requirements for good project management. According to the GAO study, this project, which has been under intense public scrutiny since 1999 and is considered as being "an essential element of the Stockpile Stewardship Program," is now estimated to cost over $4.2 billion and still has serious deficiencies in project management.

Finding. DOE continues to rely excessively on contractors for project justification and definition of scope. There are some large projects in which DOE is not effectively executing its role of owner with respect to the oversight and management of contracts and contractors.

Recommendation. DOE should develop its position as an effective owner of projects and should assure that federal project managers are trained and qualified owner's representatives, capable of dealing effectively with contractors.

PROCESS IMPROVEMENT

The Phase II report found that the process in DOE for proposing, planning, and managing projects had serious shortcomings. The remaining chapters in this report show that many of these shortcomings remain, although some steps have been taken to address them. The first step toward a process definition was taken in DOE O413.3 (DOE, 2000), but much remains to be done. A consistent, efficient, expedited, and documented process for project planning, justification, and execution is essential. The committee believes that the lack of a standard, consistent project planning process is unacceptable. It believes that significantly greater progress should have been made in the more than 2 years since the issuance of the Phase II report.

The committee recommends that DOE initiate a program of project process improvement, following an established statistical process control or continuous quality improvement procedure such as the well-known six-sigma process, which follows five steps: define, measure, analyze, improve, and control (DMAIC) (Rath & Strong, 2001). DOE has not successfully executed the first step, *define*, and has accomplished even less in the areas of *measure* and *analyze*. The six-sigma process is discussed further in Chapter 5, which gives specific recommendations. Pressures on DOE to demonstrate immediate or instantaneous improvements may be problematic because experience has shown that going directly to Step 4, *improve* (i.e., implementing preestablished solutions without passing through *define*, *measure*, and *analyze*), often leads to poor results. The committee believes that the DOE secretarial acquisition executive should be the sponsor of a program for process improvement, with OECM as the department's process improvement champion.

Finding. The DOE process for project initiation, planning, justification, and execution continues to need substantial improvement. A top-to-bottom process that recognizes best practices in both government and industry, as well as the unique and specific requirements of DOE programs and projects, is essential.

Recommendation. The DOE secretarial acquisition executive should sponsor a process improvement program, and OECM should be named the program champion in DOE.

ORGANIZATIONAL STRUCTURE AND RESPONSIBILITY

The committee finds that the project management support offices (PMSOs) in the Environmental Management (EM), Defense Programs (DP), and Office of Science (SC) program secretarial offices (PSOs) are having a positive influence on project management in their programs. Much remains to be done, but there is movement in the right direction. The committee recognizes the need for tailoring project management programs to the different PSO mission requirements. However, some PSOs are improving faster than others, and progress could be accelerated through stronger central oversight and support, using OECM as a unifying organization, to assist the PMSOs and the PSO projects and to validate the results.

The Phase II report found that project management in DOE was not being adequately addressed. To remedy this critical situation, a departmental center of excellence in project management was needed. Also, the secretary of energy and the deputy secretary/secretarial acquisition executive were in great need of a responsive organization they could rely on for accurate, unbiased project management information, advice, and early warning of problems. The Phase II report accordingly recommended the establishment of an office of project management, at the level of assistant secretary, reporting to the deputy secretary, with department-wide project management functions and responsibilities. This recommendation was reiterated in the January 2001 letter report. The committee also calls attention to the expanded functions and responsibilities of the project management office as proposed in the Phase II report.

Congress stated the following in the House Appropriations Committee report for FY2002 appropriations (U.S. Congress, 2001):

> The Department has established an Office of Engineering and Construction Management (OECM) to strengthen its project management capabilities. The Committee strongly supports this effort, but continues to be concerned with the placement of this Office in the Department's organizational structure. In its recent report to Congress, the National Research Council (NRC) reaffirmed its recommendation that the Office of Engineering and Construction Management should be at the level of assistant secretary and report directly to the Deputy Secretary. The NRC also noted that "the most important unresolved issues are: (1) definition of the authority and scope of the OECM; (2) the provision of adequate financial and staff resources to improve project management." The Committee endorses the NRC recommendation that ". . . the authority of OECM and the PMSOs be strengthened and that the resources and personnel available to them be increased to support their responsibilities." In that regard, the Committee strongly urges the Department to elevate OECM to a level equal to an Assistant Secretary with a direct reporting relationship to the Deputy Secretary/ Secretarial Acquisition Executive authority. The Committee believes that the director of this office should continue to be a career position rather than a political appointment. Further, it fully expects that OECM's existing personnel should continue in their current positions in OECM's new location. The Department should also place the facilities and infrastructure policy development and program oversight responsibilities under OECM.
>
> Consistent with NRC's recommendation for strengthening available financial and staff resources, the Committee has provided $7,600,000 for OECM in fiscal year 2002 and expects the office to report directly to the Deputy Secretary.

Finding. The combination of the OECM and the PMSOs in the three major PSOs addresses many of the issues raised in the Phase II report but not all. This organizational structure is probably workable, but it does not fully address the department-wide issues of consistency, discipline, and excellence in project management that the Phase II report felt were essential.

Recommendation. The roles and responsibilities of the OECM should be strengthened, as set forth in the Phase II report, and the OECM should be budgeted, staffed, and empowered to become the center of excellence in project management and the coordinator for project manager training and development and for oversight and approval of all capital projects in DOE.

REFERENCES

DOE (U.S. Department of Energy). 1998. Accelerating Cleanup: Paths to Closure (DOE/EM 0362). Washington, D.C.: Department of Energy.

DOE. 2000. Program and Project Management for the Acquisition of Capital Assets (Order O413.3). Washington, D.C.: Department of Energy.

GAO (U.S. General Accounting Office). 2001a. Department of Energy Status of Achieving Key Outcomes and Addressing Major Management Challenges (GAO-01-823). Washington, D.C.: General Accounting Office.

GAO. 2001b. Department of Energy: Follow-up Review of the National Ignition Facility (GAO-01-677R). Washington, D.C.: General Accounting Office.

NRC (National Research Council). 1999. Improving Project Management in the Department of Energy. Washington, D.C.: National Academy Press.

NRC. 2001. Improved Project Management in the Department of Energy. Letter report, January. Washington, D.C.: National Academy Press.

Rath & Strong, 2001. Six Sigma Pocket Guide. Lexington, Mass.: Rath & Strong/Aon Management Consulting.

U.S. Congress. 2001. Energy and Water Development Appropriations Bill, 2002 (HR 107-112). Washington, D.C.: U.S. House of Representatives.

3

Front-End Planning

INTRODUCTION

Front-end planning is, in many senses, the most critical phase of a project and the one that often gets least attention. The front-end planning process defines the project. The decisions made in this phase constrain and support all the actions downstream and often determine the ultimate success or failure of the project. Projects with adequate front-end planning do not always succeed, but those with inadequate front-end planning most often fail (CII, 1995). Typically, a project will not be better than its front-end planning process.

The front-end planning stage encompasses determination of the mission need or business objective, the scope for a project to fulfill the mission or objective, project justification, basic project definition, an outline of the general design, approximate benefits and costs, funding sources, risk factors facing the project, a basic organizational structure for the project, and a preliminary project execution plan. Based on the information developed in this phase, senior management must determine whether to approve, terminate, or modify the project. Unfortunately, this activity often takes place with insufficient attention from senior management, who often are unaware of the process and whether it has been adequately performed. Senior managers who do not spend time at the early planning stage to get a project started right will probably spend a lot of time later to fix it.

NOTE: The term "front-end planning," as used in this report, consists of the preconceptual planning phase through approval of mission need (CD-0), the conceptual design phase through approval of the preliminary baseline range (CD-1), and the preliminary design phase through approval of the performance baseline range (CD-2).

22

There are many approaches to front-end planning, which is also known as preproject planning, preconstruction planning, project programming, feasibility analysis, schematic design, scope definition, or conceptual planning. Whatever it is called, successful front-end planning requires the active involvement of senior management before decisions are made that will determine the fate of a project.

The Phase II report identified early project planning as a major factor affecting project success (NRC, 1999). It noted that inadequate definition of project scope and inadequate preconstruction planning lead to cost overruns, schedule overruns, and failure to achieve the intended project scope and performance. The report also found that adequate initial project definition was a continuing problem in DOE: "Statistical studies showed that inadequate project definition (detailed planning of scope, objectives, resources) accounts for 50 percent of the cost increases for environmental remediation projects." The report also noted that the DOE was setting project baselines too early and based on too little design information.

In October 2000, DOE issued Order O413.3, *Program and Project Management for the Acquisition of Capital Assets*, which defined the critical decision steps from CD-0 through CD-4 and a fairly detailed project-planning process as part of the capital budget cycle. OECM, in conjunction with Congress, has begun to develop a funding mechanism for project engineering and design (PED) (DOE, 2000a). As noted in the Phase II report, adequate PED funding, preconstruction planning, and project controls are all critical to successful projects. The committee reaffirms the Phase II recommendations for DOE to improve preconstruction planning and performance baselines. The committee applauds the positive steps taken by DOE for implementing preconstruction planning; however, much more management attention to improving front-end planning is needed.

An example of the need for very early project planning is the Next Linear Collider, which has been publicly proposed by the Stanford Linear Accelerator Center (SLAC), Fermilab, the Office of Science, and other laboratories. Although this project is still in the early conceptual stage, and even the country of location is undecided, the need for early front-end planning is demonstrated by the fact that cost estimates ("more than $6 billion") have already been published in the general press (Glanz, 2001a; Glanz, 2001b; Seife, 2001). It is never too early to start front-end planning, and with cost estimates having been made public, front-end planning should already be under way.

OECM has limited documentation of project planning procedures and expects to revise and expand the descriptions in the draft *Program and Project Management* manual (PPM) (DOE, 2000b) and the draft *Project Management Practices* (PMP) (DOE, 2000c), which are reviewed in Chapter 8 of this report. The committee believes that OECM and the PMSOs should seek out and implement the best, most up-to-date front-end planning methodologies. As one example, the Construction Industry Institute (CII) has defined front-end planning as "the process of developing sufficient strategic information with which owners can address risk and decide to commit resources to maximize the chance for a successful project" (CII, 1995). In this handbook, CII breaks down the front-end planning process into four steps, as shown in Figure 3-1: (1) organize for planning, (2) select project alternative(s), (3) develop a project definition package (which is the detailed scope definition of the project), and (4) decide whether to proceed with the project.

Front-end planning procedures should be focused on the process to be followed by DOE as the owner, user, and operator of the facility even though a contractor may undertake the actions. An appropriate front-end project planning process would help DOE to identify the mission need for the project and aid in identification and evaluation of alternative approaches and assessment of the costs and risks of each. It should lead to a well-defined set of requirements and scope of work that form the basis for effective design. Front-end planning in the DOE project management system includes planning procedures from project conception through approval of the performance baseline (CD-2). The DOE process includes 20 to 30 percent design completion (preliminary design) as the basis for development of a preliminary scope, budget, and schedule for the project; definition of the project performance baseline; and project funding authorization from Congress (DOE, 2000a).

ASSESSMENT OF FRONT-END PLANNING IN DOE

Past Evaluations

A recent analysis of 65 external independent reviews (EIRs) using the EM project definition rating index (EM-PDRI) included an evaluation of project planning issues common throughout DOE (RCI, 2000). Almost all the projects in the study, including many FY2000 and FY2001 projects, had significant unresolved problems. Of the 65 EIRs evaluated, 26 were missing corrective action plans. Other problems included missing or deficient cost estimates, project schedules, alternative analyses, project risk management plans, and project organization documentation, all of which are key elements in front-end planning.

Committee Assessment

The committee took steps to assess the effectiveness of the current DOE front-end planning process by requesting CD-0, CD-1, and CD-2 documents for a selected sample of capital projects in the DOE portfolio (11 projects authorized in FY2000 and FY2001, including 5 projects from DP, 5 from EM, and 1 from SC). A list of the projects and the data requested are shown in Appendix B. Unfortunately, the responses to this request were varied, incomplete, and inconsistent. The documents received did not provide the information needed for an assessment of the planning process. The amount of material on each project varied from voluminous to nearly none. Regardless of its volume, the record of

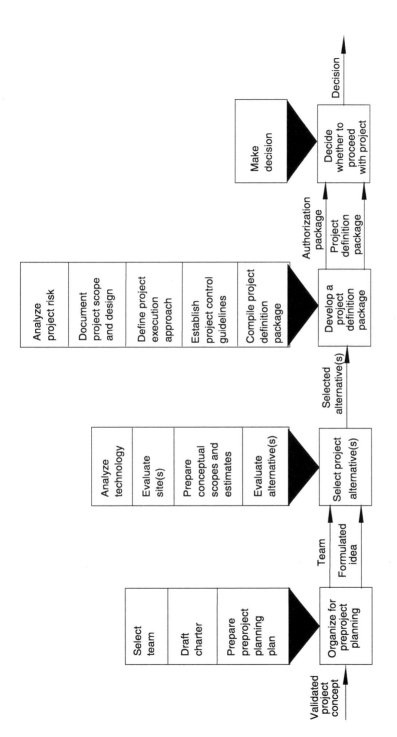

FIGURE 3-1 CII front-end planning process.

the justification of project need, definition of scope, and history of decisions was typically insufficient to permit identification and assessment of the front-end planning process. It was often impossible to determine from the documentation why a project was being performed or how its scope had been determined. Some projects did have some records of evaluations of readiness to proceed, which provided snapshots of certain steps and decision points, but based on the lack of front-end planning information in the documents that were provided, the committee determined that the current front-end planning process as applied through FY2001 is still incomplete, disorganized, inconsistent, and unreliable.

In addition, presentations on specific projects and program office policies and procedures were made to the committee at various meetings. In general, the projects were selected by the PSOs (see Appendix B for a list of presentations). From the information provided, the committee could not identify any department-wide improvement in front-end planning. It noted that while some projects appeared to be well planned and on the right track, many others demonstrated problems typically associated with inadequate front-end planning. The observed problems occurred irrespective of project size, complexity, or originating program organization.

It should be noted that all the projects reviewed by or presented to the committee were initiated prior to the activation of the OECM, the establishment of the PMSOs, and the issuance of O413.3, so that this negative assessment does not show these initiatives to be ineffective. What it does show, based on the evidence provided to the committee, is that the DOE front-end planning process for projects initiated in FY2000 and FY2001 has not noticeably improved and probably will not improve until the reforms that have been implemented have had time to become effective. Further assessments of the situation, including a review of projects funded in 2002 (when appropriate), will be made by the committee in the future; for the time being, the committee reiterates that improved project performance will require an improved front-end planning process, and that any improvement in front-end planning will require positive, aggressive action by all responsible parties.

Even though the department has issued a new policy (O413.3) and procedures (PPM manual) including front-end planning, not all program offices have incorporated these requirements into their project planning procedures. Differences in approach and attitude among the three major programs are evident.

Defense Programs (DP)—and Military Applications and Stockpile Operations (DP-20) in particular—have recognized the benefits that will accrue by following a rigorous front-end planning process and are making progress toward formalizing a multiyear implementation program. One of the more distinctive features of the DP approach is viewing the front-end planning process (through CD-2) as being a programmatic rather than a project function, thus requiring more departmental involvement. DP is trying to overcome the "wish-list syndrome" by integrating and prioritizing projects over the long term and developing budgets based on long-range plans. The committee supports the DP-20 approach and encourages its department-wide application.

EM has adapted the CII Project Definition Rating Index (PDRI) for use in its project reviews (DOE, 2001). Tools such as the EM-PDRI can help ensure consistency in front-end planning and give planners a means to assess the probability that projects will perform as planned. However, the EM-PDRI will not achieve its full potential until EM personnel are sufficiently trained in its use and it becomes an integral part of the planning process rather than an after-the-fact review tool.

The committee has observed that SC projects often involve experimental, one-of-a-kind technology, apparently leading the office to believe that documenting planning decisions consistently is not productive and that it would be impossible to devise a process appropriate for all its projects. Also, SC does not seem to have a complex-wide system for integrating and prioritizing projects for future years but validates projects as part of the annual budgetary process. By necessity, the big science projects have longer planning horizons, but nevertheless they demonstrate instances of inadequate or inappropriate front-end planning.

SC projects are typically proposed by the laboratories, individual scientists, or the research community. Projects are planned in a series of workshops in which the project scope, purpose, and research programs are developed and refined, based on inputs from the research community. Workshop participants, who are predominantly scientists and researchers, do not necessarily have recent or extensive experience with project management. The committee believes that workshops need to include project management professionals to provide support for front-end project planning, including cost and schedule estimates and risk management. Effective front-end planning should not wait for scientific consensus on scope and design.

It was noted that all program offices approach front-end planning differently for infrastructure projects and for program mission-driven projects. While smaller, less complex infrastructure projects may warrant less management attention, the components of the planning effort should be the same. It was also noted that while determining mission need should be a program office responsibility, mission need appears to emanate from the contractors and laboratories, with only perfunctory DOE oversight. The committee observes that determination and documentation of mission need are the responsibility and obligation of the owner, even when contractors perform the documentation.

Finding. Compliance with the front-end planning requirements in O413.3 has been inconsistent among PSOs and among individual projects.

Recommendation. OECM should assure that all program offices have a documented front-end planning process that meets the intent of O413.3, and that the information used as input for Energy Systems Acquisition Advisory Boards (ESAABs) and ESAAB-equivalent readiness reviews, as well as the outcomes of these reviews, is documented and used to assess project performance and progress in improving project planning.

Recommendation. The PMSOs should consider developing tailored checklists such as the EM-PDRI as in-process planning tools, train project personnel to use them, and analyze their effectiveness for projects throughout the DOE complex. Effective and consistent front-end planning should be made mandatory for all projects.

Finding. Tools such as checklists, communications software/methods, planning reviews, third-party audits, economic modeling, objective setting, and team building, if used correctly, can contribute to effective front-end planning. Performance of technical evaluation during planning is essential for projects involving new technology, complex site conditions, and complex project-flow requirements. Consistent documentation and planning structure would increase the effectiveness of front-end planning in the department.

Recommendation. OECM should clarify, expand, and revise the front-end planning procedures in the *Program and Project Management* manual and *Project Management Practices*. DOE should use standard industry procedures where applicable; however, the PMSOs should provide supporting policies and procedures tailored to the specific projects and needs of each program. The PMSOs and OECM should assure the adequacy of front-end project planning prior to each critical decision, to assure that projects are not unnecessarily delayed by poor plans and that time constraints do not cause projects to be approved without adequate planning.

Recommendation. The deputy secretary and the designated program acquisition executives should strengthen their interest and support, thereby confirming that truly effective front-end planning will be required without exception. OECM and the PMSOs should pay close attention to documentation of front-end planning decisions.

Project Engineering and Design Funding

The committee is convinced that investing in front-end planning is essential for the success of DOE project management, and it is encouraged by the creation of a funding mechanism to complete preliminary engineering and design as the basis for refined cost and schedule estimates prior to approval of the performance baseline. The committee encourages DOE to continue its development of preliminary engineering and design and other measures to define and manage risks and improve the accuracy and reliability of cost and schedule estimates.

Finding. DOE has established a process to significantly increase the accuracy and reliability of project baselines.

Recommendation. OECM should actively participate in the process and monitor the performance of projects baselined under this new process to document its impact and opportunities for improvement.

MANAGEMENT REVIEW

Front-end planning will be successful only with the involvement and support of senior management. For success, DOE senior management should insist that every project be effectively planned from its conception. Senior management should understand the process and should assure that effective project planning is being conducted. This can be accomplished by a number of means:

- Questioning at project review meetings,
- Providing resources to support the process implementation and training,
- Maintaining discipline in sticking to the plan, and
- Benchmarking results (NRC, 2001).

Because contractors are frequently the users and operators, their involvement in the front-end planning process is appropriate and necessary in most cases. However, DOE, as the project owner, has the primary responsibility for front-end planning. The committee recognizes that improvement in front-end planning cannot be incorporated uniformly in all DOE projects in a short time frame. However, consistency in front-end planning will not be achieved as long as DOE delegates this activity to contractors without also providing prescribed procedures, products, and performance measures, as well as adequate supervision. Effective front-end project planning will require both process and cultural change within the organization.

Finding. Overall, insufficient attention from DOE management is being given to the front-end planning process; however, the committee observed that management was acting in isolated cases and to varying degrees within the program offices.

Recommendation. DOE senior management should emphasize the importance of thorough and complete front-end planning (including written documentation). ESAABs and ESAAB-equivalent reviews should be used to enhance the quality of front-end project planning and assure that the project team is pursuing the right project—that is, that the project has adequate justification and will satisfy a well-conceived need.

FRONT-END PLANNING METRICS

The Phase II committee sought metrics by which to evaluate DOE's project management functions (NRC, 1999, Appendix A). Without adequate metrics, it was very difficult to evaluate the effectiveness of DOE planning practices or to compare practices among DOE projects or between DOE and private sector projects. It is equally difficult for management to address problems that exist in the diverse pool of projects that DOE performs.

For example, CII developed its project definition rating index (PDRI) to assess and guide front-end planning and related planning practices. The PDRI assesses 70 project scope-definition elements (CII, 1996, 1999). CII's *Benchmarking and Metrics Data Report* includes front-end-planning metrics derived from 23 questions in the PDRI (CII, 2000). The CII database includes over 1,000 projects representing approximately $52 billion in construction costs. The CII data show a positive correlation between front-end planning and project performance in terms of cost, schedule, change orders, and operational performance. The mean percentage of total project cost spent on front-end planning activities was 4.3 percent for the industrial projects and 2.4 percent for the building projects in CII's benchmarking database.

DOE does not currently have enough data to compare its front-end planning and project performance with best industry practices and performance. The DP PMSO has taken some positive steps to develop a benchmarking database to compare its projects with the CII database, and this positive action should be continued and extended by all PMSOs and the OECM. Also, DOE recently joined CII, so it now has access to CII's database.

Finding. Front-end planning improvement requires metrics for trend analysis. The committee was not able to obtain this information for specific projects because DOE does not have enough data for front-end planning trend analysis.

Recommendation. OECM should begin benchmarking project practices and performance metrics to identify areas in need of improvement and establish a baseline for future evaluation. This benchmarking effort should be systematic, quantitative, and analytical, and it should compare practices in industry and in other government agencies. It should capture both front-end planning and performance metrics, including actual performance versus forecast.

HUMAN RESOURCES FOR FRONT-END PLANNING

The skills needed to effectively manage the different types of DOE projects are based on technical knowledge, management experience, and personal traits. Individuals involved in science or equipment-type projects need a strong background in process engineering, mechanical engineering, or chemical engineering.

Those involved in environmental remediation projects need extensive background in environmental engineering or chemical engineering. A person perfectly at ease working on a relatively low-risk project may be lost on a project that is highly complex and changing extensively during front-end planning.

Effective planners have the technical knowledge to understand the project mission and the facilities, equipment, and processes needed to satisfy the project requirements. The project manager's experience should correspond to the level of risk and complexity of the project. In addition, project managers need personality traits that facilitate collaborative relationships. Because these skills and abilities are not easily developed after a project manager has been assigned, they should be considered as criteria for hiring and assigning personnel to planning assignments.

Finding. A training program addressing front-end planning and other project management practices is being developed. The completion date of this effort was reported to the committee to be December 2002, with training to start soon afterward. Without immediate improvement in the planning knowledge and skills of personnel and more management emphasis on improving the planning process, projects will continue to have inadequate front-end planning.

Recommendation. OECM should do more than develop policies and procedures—it should become fully engaged in process improvement beginning with front-end planning. To overcome the lack within the department of skilled project planners and the delays in training, and to bridge the gap until a training program takes effect, DOE should establish a cadre of experienced project planners within OECM; they should have a wide variety of planning capabilities and prior experience in different project types, including high-risk projects. These individuals should be a part of the initial integrated project teams and should assist the project originators (as internal consultants) in getting front-end planning done correctly, including planning prior to CD-0. This cadre of internal consultants should champion the DOE front-end planning process, providing just-in-time training for front-end planning to project teams. DOE should benchmark its management of project planning personnel and application of their expertise with that of private sector companies that have successfully undertaken similar activities. In this way, DOE may be able to jump-start an immediate improvement in planning capability.

Recommendation. DOE should eliminate impediments to initiating training for front-end project planning prior to December 2002. Training should begin as soon as possible.

REFERENCES

CII (Construction Industry Institute). 1995. Preproject Planning Handbook (Special Publication 39-2). Austin, Tex.: Construction Industry Institute.

CII. 1996. Project Definition Rating Index (PDRI)—Industrial Projects (Implementation Resource 113-2). Austin, Tex.: Construction Industry Institute.

CII. 1999. Project Definition Rating Index (PDRI)—Building Projects (Implementation Resource 155-2). Austin, Tex.: Construction Industry Institute.

CII. 2000. Benchmarking and Metrics Data Report. Austin, Tex.: Construction Industry Institute.

DOE (U.S. Department of Energy). 2000a. Program and Project Management for the Acquisition of Capital Assets (Order O413.3). Washington, D.C.: Department of Energy.

DOE. 2000b. Program and Project Management. Draft. Washington, D.C.: Department of Energy.

DOE. 2000c Project Management Practices. Draft. Washington, D.C.: Department of Energy.

DOE. 2001. Office of Environmental Management Project Definition Rating Index Manual. Washington, D.C.: Department of Energy.

Glanz, James. 2001a. "Physicists Unite, Sort of, on Next Collider." *The New York Times.* July 10, 2001, pp. D1–D2.

Glanz, James. 2001b. "To Be Young and in Search of the Higgs Boson." *The New York Times.* July 24, 2001, p. D3.

NRC (National Research Council). 1999. Improving Project Management in the Department of Energy. Washington, D.C.: National Academy Press.

NRC. 2001. Improved Project Management in the Department of Energy. Washington, D.C.: National Academy Press.

RCI (Resource Consultants, Inc.). 2000. External Independent Review Analysis. Washington, D.C.: Office of Engineering and Construction Management, Department of Energy.

Seife, Charles. 2001. "Plans for Next Big Collider Reach Critical Mass at Snowmass." *Science* 293 (5530): 582.

4

Risk Management

INTRODUCTION

Risk management is probably the most difficult aspect of project management, and for many DOE projects it is also the most critical. The Phase II report noted that DOE does not always use proven techniques for assessing, allocating, and managing risks. Discussions with DOE project managers reveal that DOE has not yet established effective risk-management methodologies or systems and that managers lack the tools and training to adequately manage risk for projects with high levels of uncertainty. The Acquisition Risk Management (ARM) pilot program by the DOE Contract Reform Office is beginning to address many risk management issues and has defined three iterative phases of risk management for DOE projects:

1. Risk identification,
2. Risk analysis and evaluation, and
3. Risk response.

The committee believes that all three phases of risk management are critical to effective project management in DOE.

The ARM process emphasizes the identification of project risks very early in the project, the development of a risk management plan during front-end project planning, and the updating of the plan throughout the project. This process, if adequately defined and implemented, has the potential to improve the mitigation and management of risks on future DOE projects. The committee supports and encourages this effort. Unfortunately, the ARM process is still in the pilot stage, and the DOE guidance on risk management issued to date is insufficient (see Chapter 8, "Documentation of Project Management Policies and Procedures").

RISK IDENTIFICATION

The proposed ARM process correctly points out that the first objective of risk analysis is to identify, define, and characterize the risks. However, simply examining the activities and work packages of a project and qualitatively assessing them as high, medium, or low risk (a process observed on several DOE projects reviewed by the committee) does not necessarily achieve the desired objectives. The disaggregation of a project into work packages, which may be very suitable for construction management and contracting, may be of little value in identifying important project risks. For example, although the performance, delivery, and cost of a critical, but yet to be developed, technology may be a major risk to project success, the technology development may not appear as a work package in a project work package analysis. Assessment of past project performance shows that risks that generated delays and cost overruns were ignored or left out in DOE work package risk analyses (NRC, 1999).

Risk factors are not only technical or environmental. Human factors—including uncertainties related to human behavior, human failures to perform, changes in critical personnel on the project or in the DOE, changes in mission or loss of mission, and other issues related to the performance of people—should be identified, quantified, and given due consideration in risk assessments.

Effective risk analysis requires an examination of the nature of the project to identify the root causes of risks and to trace these causes though the project to their consequences. As an example, if a project has a work package called *Design Advanced Superconducting Magnets Using New High-Temperature Superconducting Material*, there may be uncertainties surrounding this activity, insofar as the more advanced the design, the longer the design process may take and the more it will cost. Moreover, there may be significant uncertainties in the ultimate cost of fabrication of the equipment, the delivery date, and the ultimate performance. All of these may suggest that there are substantial risks associated with the budget, schedule, and scope of this activity. Furthermore, the activity may have even greater impact on the uncertainties associated with other work packages. In turn, the uncertainties in the physical size of the equipment, its power requirements, cooling requirements, safety requirements, maintenance requirements, reliability, and other factors may have significant impacts on still other activities during design, construction, and operation, and greatly increase the uncertainty associated with these work packages. Even so, it is not sufficient to evaluate just the primary interactions, because they, in turn, may impact other activities, and so on, making it necessary to examine the primary, secondary, tertiary impacts, or even further, to evaluate the potential for a ripple effect, in which the effects of one event propagate through many other events. Not tracking sources of risk through the causal relationships in the project can cause important project risks to be overlooked or understated. Neglect or underestimation of the ripple effect is a common deficiency in risk analysis, leading almost always to an underassessment of risk.

RISK ANALYSIS AND EVALUATION

Several approaches are available for handling the kind of risk assessment commonly associated with DOE projects:

- *Systems analysis and systems thinking.* Systems analysis and systems thinking emphasize the relationships among activities in a project and the understanding of basic feedback structures that drive projects, through the development of shared maps of the processes, participants, and their interactions.
- *Causal loop diagrams.* Causal loop diagrams are a systems analysis and thinking tool that shows how activities are related through feedback loops; they help explain why some variables have little or no effect (negative feedback) and some have highly amplified effects (positive feedback).
- *Event trees.* Event trees (or fault trees or probability trees) are commonly used in reliability studies, probabilistic risk assessments, and analysis of failure modes and effects. Each event tree shows the results of a top event and other variables, leading to the determination of outcomes and the likelihood of these outcomes.

Risk Quantification

After qualitative identification of risk factors, it is necessary to quantify them. Although there are many approaches for quantifying uncertainty, the most generally accepted methods are based on probability theory. If uncertainties are expressed as probabilities, then the entire set of methods derived from probability theory can be drawn upon. In practice, however, there are certain difficulties with their application. One problem is that probabilities are generally based on the relative frequencies of events derived from historical data, but in the absence of such data for project costs, durations, and scopes, the probabilities are not objective but subjective. A second problem arises if the person with detailed knowledge of the project is not experienced in analyzing uncertainties as probabilities. A third problem arises from the tendency of those directly involved in estimating a project's schedule or costs to underestimate the uncertainties associated with the schedule and costs. Conversely, more objective persons familiar with probabilistic methods may know little about the root causes of risks on a particular project. Collaboration and interaction between these two groups, starting from the conceptual planning stages, are typically required to produce realistic, unbiased quantification of project risks.

Many times, inputs based on subjective judgments and experience are the only information available and give useful results. However, DOE should assure that these inputs are provided by persons experienced with similar projects and that the reasons for choosing specific probability distributions are thoroughly documented. The most effective antidote to bias in risk assessments is to use an open process, with documentation, justification, and review of all assumptions by disinterested parties. Also, it is essential to the success of a risk assessment program that actual costs, schedule, and other relevant project data be collected and compared with the original estimates, in order to build up a database that can be used for estimating future projects. Without a system that provides feedback on actual performance of projects to project planners, future risk assessments will not improve.

The committee agrees with the ARM approach that emphasizes the breakdown of the uncertainties into manageable parts. Figure 4-1, presented to the committee by DOE, shows a Pareto graph of the top 40 sources of uncertainty in the River Protection Project integrated schedule. It is clear from this presentation which activities have the greatest impact on the project completion date and require the greatest attention to risk mitigation and management.

One of the important results of a good risk analysis is that it allows determining where to apply management resources and what to leave alone. Unfortunately, from the presentation in Figure 4-1 one cannot determine if the elements referenced (e.g., AZ 101 HLW Start-up HLW Vitrification Production, the highest-ranking source of uncertainty) are truly root causes or simply work packages or activities. The top events should be root causes so that this analysis can serve as a map for DOE managers to find ways to reduce, mitigate, buffer, or otherwise manage the sources of uncertainty.

Risk Modeling and Analysis (Impact Determination)

The objective of risk assessment is not just to compute risk values but to increase capacity to mitigate and manage the risks. Characterizing some risks as completely out of project management influence—acts of God, for instance—might be helpful in understanding total project uncertainty, but the primary goal of risk assessment should be the identification of active measures for risk management. Risk management should be an active not a passive endeavor.

Calculation of a single project risk estimate may be useful as input to a decision on whether to execute the project (in which case there may be biases toward underestimating it) or as a basis for setting contingency (in which case there may be biases toward overestimating it). The DOE PSOs should be in a position to know the magnitudes of risk associated with each project under their control. There are many available methods for combining risks.

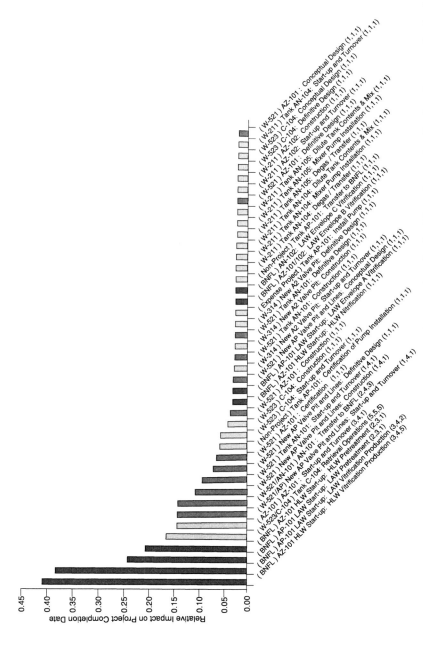

FIGURE 4-1 Pareto chart prioritizing risks: top 40 sources of uncertainty in the River Protection Project integrated schedule.

Multivariate Statistical Models (Regression Analysis)

Data-based analysis, or "objective" analysis, is one of two methods explicitly cited in Office of Management and Budget (OMB) Circular No. A-94, *Guidelines and Discount Rates for Benefit-Cost Analysis of Federal Programs* (OMB, 1992). This method is objective in that it does not rely on subjective probability distributions elicited from (possibly biased) project advocates. It builds a statistical model based on data for many projects and then compares the proposed new project with this model. Its use is highly desirable as an independent benchmark for evaluating risk (and other factors) for a specific project. Unfortunately, it requires a large database of projects, and DOE, despite its history of many projects, does not have such a database. (See NRC, 1999, Appendix B, for more information and references.)

Stochastic Simulation Models (Monte Carlo)

The Monte Carlo method is a generic term for simulations that use random number generators to draw variates from probability distributions. It is the second of two methods explicitly cited in OMB Circular No. A-94 (OMB, 1992). (See NRC, 1999, Appendix B, for more information and references.) Stochastic simulations can be very useful in the absence of real data. Their advantage is that they are based on subjective assessments of probability distributions and therefore do not depend upon large databases of project information. Their weakness is that because they are based on subjective assessments of probability distributions, their objectivity may be suspect. Stochastic simulation models that are based on event trees or feedback models can give reasonable estimates of total project risk. Monte Carlo simulations that simply add up the uncertainties associated with various activities or work packages may be biased, because the typical approach is to assume that all these activities are statistically independent. If one performs an elementary risk analysis, as described above, in which root causes of uncertainties are identified and the ripple effects tracked through the entire project, it is obvious that activities affected by the same root cause cannot be statistically independent. Therefore the inappropriate assumption of independence in Monte Carlo models can severely underestimate risks.

Simple Additive Models

If the objective is simply to find the probability distribution of the project cost estimate as the sum of a number of work package costs, stochastic simulation is unnecessary overkill. It is well known from elementary statistics that the moments of a sum are the sum of the moments, if all the terms are statistically independent. This summation needs to be modified if the variables are not statistically independent, but the summation method can still be readily applied. Simple programs based on summation of moments have been used for many years to combine risks for dependent as well as independent variables, using the first (mean), second (variance), and third (skewness) moments, permitting use of highly skewed probability distributions (project cost distributions are generally skewed to the right, that is, with long tails in the direction of higher costs). One advantage of simple additive models is that they are easily understood, and it is usually obvious which activities contribute the most to the total project uncertainty and which activities contribute relatively little. Ease of understanding is more important than a false indication of accuracy, when all the probability distributions are based on qualitative judgments anyway. Unfortunately, the summation method does not work for project durations (critical path lengths) unless one can assume, as the program evaluation and review technique (PERT) method does, that the critical path is not affected by uncertainty in the activities.

System Dynamics Models

System dynamics models are typically based on quantitative causal loop diagrams that show how activities are related through feedback loops. The models may be deterministic or probabilistic. Commercial, off-the-shelf programs are available to perform the calculations. Although systems dynamics models are more often deterministic than stochastic, because they are based on dynamic feedback principles, they can nevertheless be used to evaluate the ripple effect of various changed conditions or root causes. Experience with systems dynamics models on real projects has shown that this method generally gives much more realistic estimates of the consequences of changes or other events than methods that do not adequately account for the ripple effect. Such models are not only useful in the early stages of a project for risk assessment, they can also be very valuable in later stages for managing change.

Sensitivity Analysis

As previously stated, the primary function of risk analysis is to break down the problem into essential elements that are capable of mitigation and management throughout the life of a project. Therefore, regardless of what method of combining risks is used, it is highly desirable to perform a sensitivity analysis of the results. Not all project managers are well versed in probability theory and stochastic simulation, but experience shows that most can relate well to sensitivity analyses, which indicate the relative influence of certain variables on the outcomes. In the absence of real data, sensitivity analysis can be very useful in checking the reasonability of risk models. In fact, when a project system is tightly coupled, it may be impossible to evaluate the effect of various variables separately and in isolation; in such cases it is more effective to perform a system simulation and then use sensitivity analysis on the systems model to identify the most important sources of uncertainty.

Finding. With rare exceptions, there are no risk models for ongoing DOE projects, and back-fitting risk assessment to ongoing legacy projects does not seem to be part of the acquisition risk management (ARM) study. There is no consistent system for evaluating the relative risks of projects with respect to scope, cost, or duration, so the deputy secretary, the chief financial officer, and the PSO managers have no objective basis for knowing which projects are riskier (and therefore require more management attention) than others.

Recommendation. DOE should develop the ability to perform quantitative risk assessments. These assessments should be carried out by DOE personnel with experience in such analyses working with persons who have an in-depth understanding of a given project. Internal project risk assessments should be separately evaluated by independent assessors or reviewers who are not project proponents for reasonableness of assumptions, estimates, and results. Risk mitigation and management plans should be prepared that can deal with significant risks identified.

Recommendation. DOE project management personnel should be trained in risk assessment methodology. This training should cover not only risk analysis methodology and techniques, but also the managerial responsibilities related to interpretation of risk assessments and mitigation and management of risks.

Recommendation. Risk analyses should explicitly consider the interdependence of the various activities due to common modes (root causes), or document why there is no dependence.

Finding. DOE has not implemented statistical models (the "objective" analysis cited in OMB Circular No. A-94, *Guidelines and Discount Rates for Benefit-Cost Analysis of Federal Programs*, because it has no usable database of past and current projects.

Recommendation. DOE should develop an internal database of project data on its own projects and on projects of other owners. A system should be established to capture data on current and future projects. Data on comparable projects performed by other federal agencies and by industry should be obtained and included. The current development of the project analysis and reporting system (PARS) (discussed in Chapter 5) could be a step toward this goal, and the committee plans to follow this work with interest. Although its early stage of development prevents assessing its effectiveness at this time, the level of participation by projects, accuracy of data, completeness of data, and avoidance of duplication should be addressed by OECM. The architecture of this data system should be specifically designed to provide support for the analysis of risks for ongoing and future projects.

RISK RESPONSE

Organizational Structures and Project Uncertainties

Some risks, once identified, can be readily eliminated or reduced. Most risks are much more difficult to handle, and risk mitigation and management require long-term efforts by project managers.

If a project has a low level of uncertainty, then the optimal policy is to proceed as fast as possible. Decisions should be made as early as possible, because in a project with low uncertainty, there is by definition little chance of making bad decisions. Fixed-price contracts, perhaps with schedule performance incentives, are appropriate. In projects with cost-benefit analyses (and all projects should have cost-benefits analyses under GPRA), the present value of the project will be increased by completing the project earlier and thereby obtaining the benefits of the project sooner. The introduction of new uncertainties over time will also be minimized.

In general, everything else being equal, projects that take longer cost more. Many DOE projects take longer than they should, in part owing to dilatory decision making inside DOE and in part owing to the budgeting-authorization-appropriation cycle. Many projects seem to stall while awaiting authorization and funding and then try to make up for this lost time by rushing forward. However, if a project has a high degree of uncertainty, a full-speed-ahead approach may not be optimal or desirable. For projects with high levels of uncertainty, performance-based incentive contracts are generally more appropriate than fixed-price contracts.

In the front-end planning model in Figure 3-1, all the arrows show movement from left to right; there is no provision for rework or iteration. This may be a realistic description of conventional infrastructure projects, but rework and iteration are common in DOE projects because of factors such as design and scope changes resulting from inherent uncertainties in science, technology, and environmental characterization. Regulatory issues also provide a source of uncertainty that can cause conceptual project planning and design to be reworked many times. In high-uncertainty projects, rework is the norm, not the exception.

Failure to recognize and anticipate changes and iteration in preparing schedules and budgets can lead to unfortunate results. The use of techniques and skills that are appropriate to low-uncertainty projects can yield poor results when applied to high-uncertainty projects with great potential for changes and high sensitivity to correct decisions. For high-risk projects, a flexible decision-making approach is much more successful. Management of uncertainty cannot just be delegated to contractors; instead, attempts to assign all uncertainties to contractors have generally resulted in increased costs, unsatisfactory performance, and litigation. Effective risk management requires the active attention of federal project managers and senior program managers.

Finding. By and large, DOE's practices in risk assessment and risk management have not significantly improved since the Phase II report. The committee reviewed some project risk assessment studies but did not see an example of a risk assessment or risk mitigation plan that it finds acceptable. The discussion in the draft PPM is merely an outline, and the material in the draft PMP is not useful as a guide for practicing risk management. Conversely, the current ongoing acquisition risk management (ARM) pilot study at three DOE sites and by the Contract Reform and Privatization Office and the EM Division Steering Group/Working Group, due for completion by December 2001, is a positive move and shows promise. The committee intends to follow this study with interest as it evolves.

Recommendation. The current acquisition risk management (ARM) pilot study should be continued and expanded beyond budget risks to cover the issues addressed in the Phase II report and in this report, such as schedule, scope, quality, and performance risks.

Finding. DOE's deficiencies in risk analysis lead to inadequate risk mitigation planning and execution. Plans often address symptoms but not causes. Execution is typically reactive or nonexistent. To be useful during project implementation, this planning should, at a minimum, do the following:

- Characterize the root causes of major risks that were identified and quantified in earlier portions of the risk management process.
- Identify alternative mitigation strategies, methods, and tools for each major risk.
- Evaluate risk interaction effects.
- Identify and assign priorities to mitigation alternatives.
- Select and commit required resources to specific risk mitigation alternatives.
- Communicate planning results to all project participants for implementation.

Recommendation. DOE should develop and implement risk mitigation planning processes and standards. Project risk assessment and management should be carried out throughout the project life cycle and should be part of the documentation for each critical decision point. Risk mitigation plans should be reviewed, critiqued, returned for additional work if needed, and approved by an independent organization such as the ESAABs at each critical decision point and prior to project approval for design or construction funding.

Recommendation. Until DOE project managers can be adequately trained in risk management, OECM should establish a cadre of experienced risk assessment personnel, who can be detailed or seconded to projects in the very early stages, to provide risk assessment expertise from the beginning of projects and incorporate risk management into the initial project management plan. (Also see the recommendation for human resources in Chapter 3, "Front-End Planning.")

Strategic Flexibility

Flexibility in project plans to address foreseeable risks and flexibility in organization, management, and control to address unforeseeable events are required to successfully manage highly uncertain projects. The value of management flexibility increases in direct proportion to the uncertainty in the project. To paraphrase a quotation attributed to General Eisenhower, who said after D-day: "It is absolutely necessary to prepare battle plans, but it is equally necessary to know when to deviate from them when actually in battle." The same thought may be applied to project management. A flexible decision-making structure requires that project managers be active and show initiative. Under these circumstances, project managers should not be constrained by organizational culture, bureaucratic restrictions, fear, self-interest, or those who are likely to apply rigid management principles rather than initiative and flexibility.

Many DOE projects experience high levels of uncertainty in many critical project components. Most of these uncertainties cannot be significantly reduced through project planning alone. They require risk management approaches different from those used for traditional projects. Some of them cannot be adequately characterized and optimal actions chosen during front-end project planning. This is common when uncertainties will be reduced only over time or through the execution of some project tasks. For example, uncertainty about the presence or strength of specific chemicals in a groundwater supply or solid waste may be reduced only after project initiation and partial completion. Under these circumstances, committing to specific risk management actions during planning makes project success a gamble that the uncertainty will be resolved as assumed in planning. A classic example of commitment to a specific course of action and the maintenance of that course without developing alternative plans is the In-Tank Precipitation project at Savannah River (GAO, 1999).

Strategic flexibility can provide tools for effectively planning for and mitigating such risks. Incorporating flexibility into risk management plans can reduce project costs and durations. Flexibility can be incorporated into project planning in several ways. One approach is to subordinate the project components that are impacted by an uncertain component to the uncertain component's design. For example, if the size of a to-be-developed piece of equipment is not known, other components such as the facilities to house and service the equipment could be oversized to accommodate the 95th percentile equipment size. However this approach can generate conflicting constraints from different uncertain components, can commit the project during planning to a single action, and can be very costly in time, money, or both. Better risk management decisions can often be made after a portion of the uncertainty has been resolved through some additional testing or preliminary design of the equipment. Purposefully and strategically postponing some risk management decisions and incorporating flexibility into risk management can improve project performance. However, postponing important risk management decisions without plans for when and how those decisions will be made invites failure by allowing inappropriate reaction to short-term conditions or ad hoc decision making. Additionally, postponing decisions might be less cost effective than committing to specific actions when needed.

Finding. DOE needs to take a flexible approach in managing risk because of the high levels of uncertainty. To be effective in risk management, flexibility should be structured. A process is needed for designing, assessing, evaluating, and implementing risk-management alternatives that include decisions made during front-end project planning and decisions made after project initiation.

Recommendation. DOE should develop cutting-edge abilities to manage high-risk projects. It should adopt a process of identifying, designing, evaluating, and selecting risk management alternatives. The process should explicitly include and address alternatives that take advantage of opportunities for the partial resolution of important uncertainties after project initiation. Reviews at critical decision points should always entertain Plan B, that is, the alternatives to be pursued if the primary approach is adversely affected by subsequent information or events.

ALLOCATION OF RISK AND CONTRACTING

The committee observed that DOE in the past tried to shift risks to contractors (K.A. Chaney and J. J. Mocknick, 1998; IPA, 1993). DOE briefings, too, have implied that risks should be allocated to the parties best able to manage them, which is difficult to accomplish when DOE has no quantitative assessment of the risks. Under some circumstances, risk allocation can degenerate into attempts by project participants to shift risks to others instead of searching for equitable allocation. There are two critical starting points for risk allocation: (1) the government initially owns all the risk and the other project participants (particularly prospective contractors) own none until a contract is signed and (2) contractors generally agree to take risks only in exchange for money. There is a price that the DOE can (but not necessarily should) pay a project participant to accept the gains or losses generated by specific uncertainties, but to determine this price, it is necessary to quantify the risks. Hence, quantitative risk assessment is essential to effective contracting.

Finding. An objective assessment is essential to performance-based contracting, to assure that DOE does not shift to other project participants risks that it should retain or vice versa, or shift risks at more cost than they are worth.

Recommendation. DOE should explicitly identify all project risks to be allocated to the contractors and all those that it will retain, and these risks should be made known to prospective bidders. To use a market-based approach to allocating risks and to avoid unpleasant surprises and subsequent litigation, it is necessary that all parties to an agreement have full knowledge of the magnitude of risks and who is to bear them.

ACTIVE RISK MANAGEMENT

The management of risk during many DOE projects appears to be passive and ad hoc without the benefits of tracking the root causes of risk identified during characterization or making proactive decisions and taking actions to mitigate risks. This practice has contributed to serious project performance problems, such as at the National Ignition Facility, where some major risks were not recognized or were ignored after project initiation until the budget and schedule problems they created forced rebaselining. A passive and reactive approach is often used in which risks are generally ignored until undesired events occur, at which time solutions are sought that often assume the availability of additional resources. Such an approach precludes preventing some undesirable events and increases the costs of addressing others. Inadequate front-end risk management planning and a tradition of budget increases may be the primary contributors to these behaviors and may deter proactive risk management during projects.

Risks need to be rigorously and aggressively managed during projects. Planning for risk mitigation is an important aid for this but is not sufficient. Rigorous risk management includes monitoring every risk factor and assigning management and mitigation responsibility to project parties. One tool for this purpose is a project risk registry. This management tool is initially constructed during project planning by identifying all types of uncertainties that could impact project performance (e.g., scope, schedule, technology, permits, site conditions, and environmental) and estimating the likelihood of occurrence and the nature and magnitude of the impacts. These estimates are used for prioritizing uncertainties for managerial focus, contingency sizing, and decision making. If funds appropriated are less than requested, the project risk registry acts as a basis for rescoping or redesigning the project, so that it remains consistent with the funds allocated. After the project has started, the project risk registry provides a tool for allocating managerial responsibility for specific uncertainties and reporting and monitoring their status. The most effective use of this tool includes regular and frequent reporting on each risk until the project passes the point where the risk is no longer an issue. Risks, as the term is used here, are distinguished from work packages, which can contain risks but are not typically defined to reflect them.

The consequences of inadequate risk management were amply demonstrated by highly visible projects such as Pit 9, in-tank precipitation, and TWRS. In these examples, the risk of the initial approach not working was not adequately reassessed; they exemplify the failure to manage important risks during the project.

Finding. DOE project risks are not aggressively managed after project initiation. Risk management during projects is an inadequately developed project management capability at DOE.

Recommendation. DOE should initiate a program to improve the knowledge, skills, and abilities of project managers and develop tools and information needed to manage risk throughout the life of a project. Project participants who manage risks actively and achieve successful project performance should be appropriately rewarded.

ONGOING PROJECT RISKS

The committee observed that many ongoing DOE projects are characterized by a high level of uncertainty and a minimal understanding and management of their risks. Most of these projects were initiated before the Phase II report and before DOE initiated project management reforms. Further, the committee observed a deficiency in DOE risk management methods. DOE has a need to manage risks to project schedules, cost, and scope. Doing so would prevent unpleasant surprises, enable remedial action, and avoid breaching baselines.

The committee believes that DOE should conduct a risk analysis of all ongoing large projects to establish their risks and vulnerabilities with respect to schedule, cost, and performance. The analysis could be used to establish a department-wide assessment of the risks remaining in each project and for the department as a whole, as well as to identify projects that are the most vulnerable and need the most attention. Because consistency is necessary for department-wide, cross-project comparisons, it is recommended that this risk study be led by OECM.

Finding. The committee observed an ongoing deficiency in risk management that undermines DOE's ability to avoid surprises and take timely remedial action to avoid baseline breaches and to predict the actual cost to complete ongoing projects.

Recommendation. DOE should conduct an immediate and thorough risk assessment of all ongoing DOE projects with significant remaining time and costs. Such an assessment would establish, on a consistent basis, the risks and vulnerabilities of projects with respect to schedule, cost, and performance. It should assess the actual status of current projects and compare them with the project's original baselines, the current project schedules and budgets, and performance for comparable completed projects. The assessment should evaluate the risks of future scope shortfalls and budget and schedule overruns.

DEVELOPMENT OF RISK MANAGEMENT EXCELLENCE

It is not unusual for DOE projects to have unusually high levels of uncertainty in many critical project components. The successful management of these risks is often critical to project success. However, traditional risk management tools, methods, and practices may be inadequate. The committee believes that the methods in the draft PPM and PMP documents are inadequate if applied in a piecemeal fashion to the task of assuring successful project management practices under the conditions pervading DOE projects. Given the circumstances, new risk management tools and methods should be developed, tested, and implemented within DOE. Several existing tools and methods, such as those cited earlier and below, and the successful (or unsuccessful) management of risks in engineering projects that match DOE projects in size and duration (Miller and Lessard, 2000) can guide this effort and form the basis for developing risk management excellence at DOE. DOE has funded the development of a number of risk models (Diekmann, 1996; Parnell et al., 1997; Diekmann and Featherman, 1998), but there is little evidence that they have been used on actual projects.

The DOE could set up a project simulation program that would let project managers simulate the activities in a project before doing it. This was done successfully in private industry as well as in the military. Simulation could be manual or computerized or both. A project simulation facility might be expensive, but it would certainly be less expensive than making big mistakes on real projects and would pay for itself in the long run. Computer simulation models have been used to study the feedback loops and the effects of change in projects, and they have been used successfully to describe and to predict project completion rates and costs.

High-risk projects are not well described by conventional critical-path network models (which prohibit recycling), and efforts to apply conventional methods inappropriately to these projects can lead to incorrect conclusions and counterproductive solutions. One approach to developing useful computer project simulations is system dynamics. This computer simulation modeling methodology can specifically depict the characteristics of dependencies among project processes, resources, and management and their impact on project performance. By focusing on specific issues, modeling can clarify and test the assumptions used by project participants and be used to design and test existing and proposed project process improvements and managerial policies.

Finding. Innovative, cutting-edge, and exceptional risk management abilities are needed by DOE to identify and address the risks in many of its projects. DOE needs to develop expertise and excellence in managing very risky development projects. The DOE complex has the intellectual, computational, and other resources necessary to produce significant improvements in this area.

Recommendation. DOE should develop more expertise and improved tools for risk management. Nontraditional and innovative approaches, tools, and methods should be investigated for their adaptability to DOE project conditions and use in DOE risk management. They would include those cited earlier in this report and in the Phase II report (NRC, 1999, Appendix B), such as systems analysis, event trees, causal loop diagrams, system dynamics, and stochastic simulation, which have been tested and shown to be valuable on similar projects or in addressing similar challenges.

PROGRAM RISKS ACROSS MULTIPLE PROJECTS

The discussion above has addressed risks mainly at the individual project level. However, of at least equal concern is the management of risks at the PSO level and at the departmental CFO or AE level. It is often said that project budgets and contingencies should be based on risk assessments, that is, on probabilities. Although probabilistic statements are impossible to verify on the basis of a single observation, DOE performs a large number of projects, so that statistical statements could in principle be verified over the population of all projects. The following is an elementary example.

The committee has been informed that the appropriate level of authorization for a project, assuming that the uncertainty in the ultimate project cost can be described by a probability distribution, is some value that has been called the risk-adjusted cost estimate (RACE). This might be, as one example, the dollar value at, say, the 85th percentile or confidence level. That is, using this number, there would be an 85 percent probability that the project will actually cost less than the RACE and a 15 percent probability that it will cost more. So, if there are 100 such projects, all funded at their respective RACEs, one would expect that 85 of these projects would be completed within their budgets and 15 would return to Congress for additional funds. In other words, if budgets are set at the RACE of the 85th percentile, statistically, 85 percent of the projects should return some unused funds to the treasury. This does not appear to be the case. Of the projects presented to the committee, only one, a DP project at Los Alamos (Infrastructure Renewal—Water Well Replacement, funded at approximately $17 million), claimed to have given back the contingency. No systematic data were available to the committee, as DOE does not seem to track contingency funds or management reserves. The conclusion, therefore, is that whatever the budget allocated to a project, the project will rarely or never underspend this budget, although it might overspend it.

For there to be any accountability, not to mention management of contingency funds, it would be necessary to state who authorizes the transfer of contingency funds to the baseline. DOE policy and procedures should define whether this is the federal project manager, the change control board, the ESAAB, the CFO, the contracting officer, or some other entity. DOE documentation should define policies when more than one PSO, or multiple offices or laboratories, are involved in a single project—that is, whether each laboratory controls (i.e., is able to reallocate) its own management reserve or whether the management reserve on a multilaboratory project should be under the control of the overall project manager, controlled by the DOE site office manager, or controlled at the PSO level.

Contingency in the schedule is as important as contingency in the budget and should be also be covered by DOE policy. None of the above policy issues are covered in O413.3 or in the draft PPM manual or the draft PMP.

Finding. DOE does not seem to have a consistent or explicit policy on the use of management reserves, what size they should be, and who should control them.

Recommendation. The deputy secretary as secretarial acquisition executive, and the chief financial officer, assisted by the PSOs and OECM, should define and state DOE policy on management reserves. This policy should be clarified in a future release of O413.3.

REFERENCES

Chaney, K.A., and J.J. Mocknick. 1998. Privatization: A Business Strategy Under Siege. WM 98 Proceedings, Tucson, Ariz.

Diekmann, J.E. 1996. Cost Risk Analysis for U.S. Department of Energy Environmental Restoration Projects. A Report to the Center for Risk Management, Oak Ridge National Laboratory. Boulder, Colo.: University of Colorado Construction Research Series.

Diekmann, J.E., and W.D. Featherman. 1998. "Assessing Cost Risk Uncertainty: Lessons from Environmental Restoration Projects." *Journal of Construction Engineering and Management* 124 (6): 445-451.

GAO (General Accounting Office). 1999. Process to Remove Radioactive Waste from Savannah River Tanks Fails to Work (GAO/RCED-99-69). Washington, D.C.: General Accounting Office.

IPA (Independent Project Analysis). 1993. Project Performance Study for U.S. Department of Energy, Office of Environmental Restoration and Waste Management. Reston, Va.: Independent Project Analysis.

Miller, Roger, and Donald Lessard. 2000. The Strategic Management of Large Engineering Projects. Cambridge, Mass.: MIT Press.

NRC (National Research Council). 1999. Improving Project Management in the Department of Energy. Washington, D.C.: National Academy Press.

OMB (Office of Management and Budget). 1992. Guidelines and Discount Rates for Benefit-Cost Analysis of Federal Programs (Circular No. A-94). Washington, D.C.: Executive Office of the President.

Parnell, G.S., J.A. Jackson, J.M.Kloeber, Jr., and R.F. Deckro. 1997. Improving DOE Environmental Management Using CERCLA-Based Decision Analysis for Remedial Alternative Evaluation in the RI/FS Process. Report VCU-MAS-97-2. Butte, Mont.: MSE Technology Applications.

5

Project Reporting and Oversight

INTRODUCTION

The Phase II report (NRC, 1999) concluded that DOE has no acceptable financial and project reporting system and recommended that DOE establish such a system to provide the data necessary for each level of management to track the cost, schedule, and scope of all projects and to support the statistical analyses necessary for process improvement. Each PSO has its own active project reporting system, and OECM has completed a specification for a department-wide project analysis and reporting system (PARS). A beta version of PARS was released in June 2000 and is being reviewed and improved. The committee supports this initial step but is concerned that the system does not include analysis or automatic data collection tools. The committee believes that the PARS should be designed so that it supports the data analysis needed by project managers to evaluate project performance as well as the oversight needs of the PSOs, OECM, CFO, and the deputy secretary. The database should also provide the information needed for benchmarking future projects.

Effective oversight of project performance is dependent on the systematic and realistic reporting of project performance data. The committee has observed some resistance in the field to changing or adding reporting practices to conform to a uniform system. Clearly, organizations within DOE have become accustomed to different types of reports that satisfy their needs, and these legacy systems should be taken into consideration when designing a new reporting system. Successful implementation of any management information system requires knowledge and consideration of the needs and preferences of the users. A critical factor in the success of a reporting system is that those who are burdened with the input of data should also receive some benefits from doing so in an accurate and timely manner. For quality and consistency, it is necessary that each data element be input only once, as close to the source as possible. A schedule to phase in reporting requirements in a manner that does not disrupt ongoing projects or cause unnecessary costs may be needed. Reporting cost data on an accrual accounting basis, rather than a cash basis, is critical for tracking and managing planned and actual cost information in any management information system. At present, accrued cost information is not generally available throughout DOE.

ANALYSIS OF EARNED VALUE MANAGEMENT SYSTEM DATA

OECM has reported that PARS will display earned value management system (EVMS) data on projects, but the accrued cost accounting system to support EVMS reporting must first be implemented. The OECM presentation indicated that project oversight is to be accomplished by plotting the budgeted cost of work scheduled (BCWS), the budgeted cost of work performed (BCWP), and the actual cost of work performed (ACWP) versus time. The example used in the OECM briefing is for the Tritium Extraction Facility (TEF) at the Savannah River site. The plot of BCWS, BCWP, and ACWP versus time for this project is reproduced here as Figure 5-1. These conventional forms of data presentation

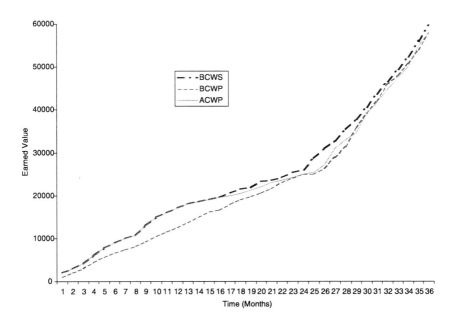

FIGURE 5-1 Cumulative earned value data for the Tritium Extraction Facility.

show the cumulative BCWP and ACWP, which tend to obscure the magnitude of current changes. For all but the very early phases of a project, the cumulative BCWP and ACWP numbers are largely determined by project history and very little by current events. It would take a substantial change in any single reporting period to have any visible effect on the cumulative BCWP and ACWP. Moreover, any differences from period to period are shown only as changes in slope, which are difficult to see. The plots serve to show whether the ACWP curve is consistently over or under the BCWP curve, and the BCWP curve over or under the BCWS curve, but little else. The plot in Figure 5-1 shows that it is easy to discern long-term trends after they have happened (for example, the BCWP is well below the BCWS in the first year or so) but very difficult to discern what is happening currently, owing to the necessary scale of the plot and the inertial effect of past history.

Two nondimensional indexes, the schedule performance index (SPI = BCWP/BCWS) and the cost performance index (CPI = BCWP/ACWP), can also be readily plotted. Current performance, as gauged by these indexes, is also obscured by the inertia of past history. Examination of the SPI and CPI values for a number of DOE projects indicates that almost all of them habitually fall between 0.95 and 1.05. In many cases, it seems that these results are due to continual rebaselining of the project budgets, so that the reports can show CPIs of approximately 1.0. As one example, the SPI and CPI values for the National Ignition Facility (NIF) project were reported to the committee in February 2001 as very close to 1.0. This shows that as long as there is rebaselining, through properly processed change control actions, the reported SPIs and CPIs provide little information and in fact may be misleading. The contingency utilization index, discussed below, would reflect the change as a debit to the contingency.

PROJECT OVERSIGHT

DOE management needs to be able to detect potentially adverse trends in project progress and to distinguish them from mere random fluctuations in reporting. Senior management needs data to decide when intervention is necessary to correct an adverse trend and when no intervention is needed.

One approach to meeting these two needs is to extract more useful and up-to-date information from the EVMS data and to analyze it using the well-known methods for statistical process control (SPC) charts. SPC (or SQC, statistical quality control) has been used in the manufacturing industries for at least 70 years (Shewhart, 1931) and was the driving force behind the rebuilding of the Japanese manufacturing industry in the 1960s. The use of control charts for the analysis of EVMS data is discussed briefly in Appendix E.

CONTINGENCY UTILIZATION INDEX

Because many DOE projects involve first-of-a-kind or one-of-a-kind applications of technologies and/or project teams not necessarily accustomed to the management of complex projects or R&D programs, it is recommended that DOE incorporate an additional index into its EVMS reporting system. This is the contingency utilization index (CUI). The CUI is defined as the contingency remaining at any point in time divided by the estimated cost to complete at that time. At the start of the project, before any work is done, this index reduces to the assigned project contingency, or risk-adjusted cost estimate (RACE) less the budget at completion (BAC) divided by the budget at completion. A formula that can be used at any time t and that reduces to CUI = (RACE – BAC)/BAC at the start, when $t = 0$, is the following:

$$CUI = [RACE - BAC + BCWP(t) - ACWP(t)]/[BAC/CPI(t) - ACWP(t)]$$

where RACE – BAC = the original contingency; $ACWP(t) - BCWP(t)$ = excess (if any) of actual cost over budgeted cost (i.e., contingency used up); $BAC/CPI(t) - ACWP(t)$ = estimate to complete; $CPI(0) = 1$; and $ACWP(0) = BCWP(0) = 0$.

As the work progresses, the CUI is adjusted to reflect the status at each period, and it can be plotted to compare the planned CUI over the period of the project with the actual values reported. In addition to stating the current CUI, each period report should state the purposes for which the contingency has been utilized. If the project accounting is done on an accrual basis, as recommended, this can be timely information, even preceding the completion of pending change procedures. If the information is not timely it delays recognition of important project budget and contingency utilization data and skews project performance information being reported.

The CUI should be reported in addition to the more standard measures CPI and SPI, as it is quite possible, for example, that the CUI is increasing (and therefore the contingency looks good) at the same time that the CPI is decreasing (costs are exceeding budget). Likewise, when the BCWS is adjusted through a change control action, the CUI will reflect the debiting of the contingency, thus alerting managers. The depletion of contingency might be missed if only CPI and SPI are reported.

BENCHMARKING

The Phase II report (NRC, 1999) included the recommendation that DOE should undertake project performance studies of all major construction projects and use this information to benchmark project performance and measure progress in project management. The committee had great difficulty in locating any documentation of project decisions, project management processes, and project performance. The PSOs have quarterly project reporting systems in place, but they provide only summary information that is insufficient for benchmarking. The committee supports the DP initiative to participate in the annual CII project management benchmarking survey and to use the CII database to identify best practices within DOE and to compare DOE performance with that of private industry. However, all DOE projects should be benchmarked, and this benchmarking should be consistent across the department. It may not be efficient for every PSO to repeat this process independently.

MEASURING IMPROVEMENT IN PROJECT MANAGEMENT

The committee has been seeking metrics for measuring DOE improvement in project management and the efficacy of committee recommendations. The evaluation of project outcomes is an obvious metric, but this can be done only after projects are complete. Moreover, as noted elsewhere, this evaluation is made difficult because DOE often rebaselines projects. That is, one might look at the actual dates and costs at project completion, assuming the data are not adjusted by rebaselining, but virtually any project that completes during the committee's tenure will have been started before the Phase II report. Conversely, projects that start after the date of this report will not finish until long after the committee's term expires. Therefore, some short-term metrics are needed.

One possible metric is to examine the means and variances in the schedule performance index for a given reporting period ($SPI(t)$) and the cost performance index for a given reporting period ($CPI(t)$) for all DOE projects (see Appendix E). Because $SPI(t)$ and $CPI(t)$ are normalized dimensionless ratios, they are not influenced by project size or duration. Therefore, all projects can be compared on the same basis. The band between the upper and lower process limits can be considered a measure of the quality of DOE project management. If this band becomes smaller over time, then DOE project management is improving. If reasonable specification limits are set and the six-sigma process limits lie inside the specification limits, then one could say that DOE project management has achieved six-sigma quality.

OTHER METRICS

One advantage of the $SPI(t)$ and $CPI(t)$ run plots, as discussed above, is that the data should be readily available through the EVMS project reporting systems; no new data sources are required. Many other metrics are possible. One area that the committee specifically recommends for measurement is that of changes: engineering change requests, disposition forms, and other documentation should be standardized and reporting metrics defined so that change can be quantified. As previously discussed, the measurements should include performance and practice use metrics. Additional metrics should be identified and defined.

FINDINGS AND RECOMMENDATIONS

Finding. The committee has observed some objections to changing or adding reporting practices to conform to a uniform system. Clearly, each organization in DOE has become accustomed to its own reporting system, and these legacy systems should be taken into consideration when designing a new department-wide reporting system. Successful implementation of a management information system requires a knowledge and consideration of the needs and preferences of the users. A critical factor in the success of a reporting system is that those who are burdened with the input of data should also receive some benefits from doing so in an accurate and timely manner. For quality and consistency, it is necessary that each data element be input only once, as close to the source as possible. A schedule to phase in reporting requirements in a manner that does not disrupt ongoing projects or cause unnecessary costs may be needed.

Recommendation. DOE and its contractors should adopt full accrual cost accounting systems in order to provide EVMS and PARS with appropriate data.

Recommendation. The PARS information system for collecting data from projects department-wide should be designed so that it supports the data needs of project managers to evaluate project performance as well as the oversight needs of the PSOs, the OECM, the CFO, and the deputy secretary. The database should also provide information for benchmarking future projects.

Finding. DOE management needs to be able to detect potentially adverse trends in project progress and distinguish them from mere random fluctuations in progress reporting. EVMS data provide some very valuable insights into the health of a project and can predict the probable outcome. They can also shed light on the conduct of the work, particularly when it is reported and analyzed to evaluate period-to-period trends.

Recommendation. DOE should utilize EVMS data to calculate the incremental and cumulative cost performance index (CPI), schedule performance index (SPI), and contingency utilization index (CUI) for each reporting period to analyze and improve project performance.

Finding. The committee had great difficulty in locating information documenting project decisions, the project management process, and project performance. DP is planning to participate in the CII benchmarking survey, but there is generally not enough consistent information to allow benchmarking project management performance within DOE or between DOE and other federal agencies and private industry.

Recommendation. All DOE projects should be benchmarked within DOE and between DOE and other federal agencies and private industry, and this benchmarking should be consistent across the department.

REFERENCES

NRC (National Research Council). 1999. Improving Project Management in the Department of Energy. Washington, D.C.: National Academy Press.

Shewhart, W.A. 1931. Economic Control of Quality Manufactured Product. New York, N.Y.: Van Nostrand.

6

Independent Reviews

INTRODUCTION

The committee finds that DOE has made substantial progress in the implementation of reviews and the resultant corrective action plans and in the formalization and institutionalization of the review process since the issuance of the Phase I report (NRC, 1998), the Phase II report (NRC, 1999), and the NRC letter report dated January 17, 2001 (NRC, 2001). The committee continues to recommend the use of formalized assessments of management, scope, cost, and schedule at appropriate stages—from determining project need to determining readiness for construction—as well as regular performance reviews. The committee also recognizes that the Energy and Water Subcommittee of the House Appropriations Committee continues to rely heavily on external independent reviews and has mandated that all line-item projects be reviewed before any new money is spent. Despite the added emphasis on reviews, the potential for cost overruns for some projects, such as the National Ignition Facility, the Isotope Production Facility, and the Neutrinos at the Main Injection Facility, continue to evoke concern. A June 2001 GAO report notes that "NIF still lacks an independent external review process" (GAO, 2001). New projects just getting under way are also of great concern to the committee. Timely reviews in the front-end project planning stage could be very helpful in getting new projects off to the right start.

The committee heard some DOE field representatives at various forums endorse the reviews and was encouraged by these endorsements. At the same time, other individuals said they had problems with the required scope and detail of the reviews, their number and frequency, and their application to small projects.

This may suggest that the procedures and general requirements should be more carefully tailored to the circumstances of the different projects to assure that the reviews are cost effective, as recommended in the Phase II report.

DOCUMENTATION OF REVIEW PROCEDURES

The committee reviewed three documents—the Office of Defense Programs (DP) Project Review Procedures, dated September 19, 2000 (DOE, 2000a); the Office of Environmental Management (EM) Internal Independent Review Handbook, dated August 2000 (DOE, 2000b); and the Office of Science (SC) Independent Review Handbook, dated January 2001 (DOE, 2001a)—which describe the respective PSO review procedures for internal independent reviews. The DP and EM documents are formatted differently, but both are fairly comprehensive and detailed and appear to satisfactorily cover the aspects expected of a review. Both focus on reviews preceding a particular decision point. A distinctive feature of the Environmental Management approach is the use of the EM project definition rating index (EM-PDRI), which assesses how well a project is defined in terms of a numerical score. EM based its index on rating factors developed by the CII. EM-PDRI scores are an important element in the decision to proceed to the next phase, but are only one of several elements in a go/no-go decision (DOE, 2001b).

Although their approaches differ, DP and EM are to be commended for aggressively addressing this issue. By contrast, the SC handbook is less formal and detailed, perhaps because a process for review of scientific and technical issues has long been institutionalized at SC. The SC reviews are not specifically oriented to the critical decision points defined in DOE O413.3. They serve to validate technical, cost, and schedule baselines before construction funds are requested. Most SC reviews are conducted semiannually to assess the status of ongoing projects. These characteristics of SC reviews were commented on in the Phase I and Phase II reports (NRC, 1998, 1999).

The committee also reviewed the Office of Engineering and Construction Management (OECM) draft Independent Review Procedure, dated June 1, 2001 (DOE, 2001c), which addresses both internal and external reviews department-wide. This document most closely resembles the handbook issued by DP. The committee finds the OECM document to be comprehensive and well presented. A department-wide manual to achieve consistency in process, nomenclature, and reporting is strongly recommended by the committee. Department-wide procedures would mean broader recognition for the lessons learned in reviews and would facilitate the ability of reviewers to participate in reviews in multiple organizations. Any distinct features of a particular program could be identified and articulated in the text or in appendixes.

The OECM document names OECM as responsible for executing the external independent review (EIR) and the independent cost estimate (ICE) and names the respective PSOs as responsible for executing the internal project review (IPR) and the internal cost review (ICR). The Phase II report (NRC, 1999) recommended that internal reviews be managed centrally; however, the committee is satisfied that by following the proposed OECM procedures, internal reviews can be executed by the respective program offices with OECM oversight.

REVIEW TEAM QUALIFICATIONS

Criteria for members of teams carrying out internal independent reviews vary among the PMSOs and OECM. SC requires the reviews to be conducted by a nonproponent review team and defines "nonproponent" as reviewers having no current affiliation with the project being reviewed and not being from the responsible program office in SC, related contractors, or the related funding office. The SC team leader is always a DOE employee. EM stipulates that reviewers have no current affiliation with the project and no current assignment to an organization participating in the project. EM team leaders can have neither current nor prior involvement with the project, nor can they have current affiliation with a participating line program or field site. The members of the DP review teams must be selected mainly from outside DP and from federal agencies outside DOE. The OECM procedure follows the DP concept, which is most reliable for ensuring that reviewers are truly independent and have no conflict of interest. The committee supports this approach but recognizes that it could become burdensome and expensive for small and/or uncomplicated projects.

REVIEW REQUIREMENTS

DOE O413.3 specifies an external independent review (EIR) of the performance baseline prior to CD-2 and an internal project review (IPR) of readiness for construction prior to CD-3 for all projects with a total project cost (TPC) exceeding $5 million. If the project is classified as a major system (above $400 million), an internal project review is required prior to CD-0 and an EIR prior to CD-3.

The committee is concerned that mandatory review for projects between $5 million and $20 million TPC may be consuming too many resources and diverting too much management attention relative to the value added. The Phase I report recommended lowering the floor from $20 million to $5 million total estimated cost (TEC) only if certain criteria prevail. A number of DOE field representatives have voiced objections to the time and manpower involved in supporting the reviews for minor projects. The committee understands that although O413.3 refers to the $5 million level, OECM intends to waive the requirement if a review would not be cost effective. In any case, there is nothing to preclude an acquisition executive or other authorized DOE executive from ordering a review for any stage of any project without requiring reviews for all projects greater than $5 million.

An area of concern during the Phase I and Phase II studies was the weakness of DOE's level of effort in front-end project planning and its lack of documentation for such planning. As noted above, the only formal review requirement is an IPR prior to CD-0 for major systems. The committee reviewed the subject further with the three PSOs and found that DP and EM recognize the weakness and are devising a means to strengthen the process. The committee believes much could be gained by expanding the use of independent reviews in the front-end project planning phase. It would appear appropriate to require an IPR prior to CD-0 for projects less than $20 million and require an ICR prior to CD-1 for projects over $20 million.

REVIEW EVALUATION

External Reviews

An external independent review process for projects was initiated by DOE subsequent to the issuance of the Phase I report, which recommended a number of specific project reviews (NRC, 1998). As time was of the essence to meet congressional mandates, DOE awarded contracts to various engineering firms and other consultants to perform the reviews. In some cases, the haste probably contributed to ineffective reviews—there may have been inadequate scope definition and/or selection of contractors that were underqualified or not truly independent. Nevertheless, the reviews exposed many defects in project definition, and the cost and schedule estimates and reviews were, on balance, a valuable tool for project management and maintaining credibility with the Congress.

The inconsistency in content, quality, and format of the early review reports, plus the fact that some reports were not well received internally or externally, illustrates the need for a more consistent and centralized approach and control. The committee strongly supports the decision to assign the responsibility for EIRs to the OECM. OECM has taken action to define the scope of an EIR more specifically in order to obtain more useful and consistent results and to decrease the time and money required to conduct an EIR. The continuing improvement of EIR reports should help project managers increase the probability of project success and better assure Congress of the credibility of project performance reports.

OECM contracted for a statistical analysis of 65 EIRs transmitted to Congress through fiscal year 2000 (RCI, 2000). The data were collected and organized around the lines of inquiry of the EM-PDRI system. Of the projects reviewed, 94 percent identified issues related to management, planning, and controls; 84 percent identified issues related to cost; 83 percent identified issues related to scope and technical factors; 66 percent identified issues related to schedule; and 23 percent identified issues related to external factors. The issues identified in reviews strongly indicate that a large majority of projects were not ready to proceed to the next stage without corrective actions. Figures of such magnitude clearly portray the value of reviews as well as the desirability of continuing the EIR program until positive improvement is demonstrated.

Internal Reviews

The committee evaluated a sample of internal review reports from DP, EM, and SC. It commends the reports from DP for their penetration of the issues and uniformity of presentation. These reports will provide an excellent audit trail for future reviews and critical decisions. The reports from EM were less detailed because the use of the EM-PDRI allowed the narrative to be condensed. There was no attempt to judge the relative value of the EM and DP approaches, except that the DP reports appear to represent a larger investment of time and money. It is noted that, on occasion, DP performed an internal review as preparation for defining the contract for an external review. These internal reviews prior to external reviews may be very valuable, but they should be recognized as part of the cost of the review program. The SC report reviewed the status of project technical performance and management activities and contained less analysis of management procedures.

PROJECT REVIEW MANAGEMENT SYSTEM

OECM recently developed a computerized project review management system (PRMS) as a tool for improving the quality of project reviews. The system stores pertinent project cost and schedule data, tracks progress, stores reports, schedules reviews and follow-up actions, creates statement-of-work templates, and promotes consistent report formats. The system also contains a master list of questions (lines of inquiry) to guide the review. The questions were developed from the EM-PDRI and DP methodology. Findings and recommendations emanating from the reviews are transferred to a corrective action plan (CAP), which is monitored through completion. The committee endorses the automation of the project review process and the tracking of the ensuing corrective actions but recognizes that the system is evolving and will be adjusted as experience is gained.

FINDINGS AND RECOMMENDATIONS

Finding. The evidence available to the committee indicates that the EIR program continues to identify significant management issues in the projects reviewed and in DOE's operation in general. Absent substantial evidence of improvement in DOE project management and project performance, the EIR program needs to be continued.

Recommendation. DOE, through OECM, should establish performance metrics for the EIR program that identify trends and opportunities for improving project management performance.

Recommendation. The EIR program should continue in its present form under OECM direction until there is clear evidence of improvement in DOE project management and project performance.

Finding. DOE would benefit from a department-wide procedure governing external and internal independent reviews. Consistent procedures would increase the pool of qualified reviewers, expedite the review and report process, and enable an automated system for tracking deficiencies and corrective actions.

Recommendation. DOE should expedite the issuance of the Independent Review Procedure drafted by OECM.

Finding. A more thorough review analysis for defining mission need and setting a preliminary baseline range during the front-end project planning phase would give the decision makers more useful information.

Recommendation. DOE should expand the use of IPRs for the CD-0 decision and should require an ICR prior to CD-1.

Finding. There is some concern that mandating formalized reviews for projects costing between $5 million and $20 million TPC may be dedicating manpower and money beyond the point of significant value added. It also may be distracting project personnel from doing the project and diverting DOE's project management resources from larger, more complex projects.

Recommendation. DOE should reevaluate the benefits gained from mandating reviews for projects costing between $5 million and $20 million TPC. The OECM should establish guidelines to evaluate the cost-effectiveness of review. At a minimum, the $5 million threshold should be based on TEC and provide for significant tailoring of the review process.

Finding. The EM Project Definition Rating Index (EM-PDRI) analyzes the readiness of a project by rating it on a numerical basis. It allows making judgments based on a multitude of rating factors, including risk, but users will need training and experience with the index in order to achieve uniformity of application and confidence in the results.

Recommendation. DOE should explore the potential application to other programs of the PDRI approach adopted by EM.

REFERENCES

DOE (U.S. Department of Energy). 2000a. Office of Defense Programs Project Review Procedures. Washington, D.C.: Department of Energy.

DOE. 2000b. Office of Environmental Management Internal Independent Review Handbook. Washington, D.C.: Department of Energy.

DOE. 2001a. Office of Science Independent Review Handbook. Washington, D.C.: Department of Energy.

DOE. 2001b. Office of Environmental Management Project Definition Rating Index. Washington, D.C.: Department of Energy.

DOE. 2001c. Office of Engineering and Construction Management Draft Independent Review Procedures. Washington, D.C.: Department of Energy.

GAO (General Accounting Office). 2001. Department of Energy: Follow-up Review of the National Ignition Facility (GAO-01-677R). Washington, D.C.: General Accounting Office.

NRC (National Research Council). 1998. Assessing the Need for Independent Project Reviews in the Department of Energy. Washington, D.C.: National Academy Press.

NRC. 1999. Improving Project Management in the Department of Energy. Washington, D.C.: National Academy Press.

NRC. 2001. Improved Project Management in the Department of Energy. Washington, D.C.: National Academy Press.

RCI (Resource Consultants, Inc.). 2000. External Independent Review Analysis. Report for Office of Engineering and Construction Management, Department of Energy, November 30.

7

Acquisition and Contracting

INTRODUCTION

The committee reiterates the finding that acquisition planning and acquisition and contracting techniques play an integral role in successful project management (NRC, 1998, 1999). This role is particularly critical at DOE, where approximately 90 percent of the department's budget is expended by contract (GAO, 2001). The findings and recommendations of the Phase II report stress the importance of developing and employing contracting methods that ensure accountability, adequately address risk, and focus the government and the contractor on achieving the outcomes sought. The committee continues to advocate these recommendations.

In accomplishing its mission, DOE relies on different types of contractor-managed activities. They include the large national laboratories, generally university-run or consortium-run, which serve as federally funded R&D centers for critical nuclear weapons and their design, development, and stockpile stewardship efforts. These activities also include large private sector management and operation (M&O) and management and integration (M&I) contractors that maintain operations at sites such as the Nevada Test Site or perform cleanups of hazardous and radioactive waste at places such as Rocky Flats, Colorado.

Under an M&O contract, the contractor/subcontractor team performs much of the work under a cost-plus, award-fee contracting arrangement. Under an M&I contract, the contractor conducts continuing competitions among subcontractors and awards fixed-price tasks where possible to get the most cost-effective solution. DOE also uses contractor support for other purposes, such as ammunition assembly at the Pantex plant in Texas.

PERFORMANCE-BASED CONTRACTING

Over the past decade, a number of contracting reforms have been put in place across the government to streamline and simplify processes and to sharpen the focus on agency mission and results. Performance-based contracting (PBC), promulgated as a government-wide policy in 1991 by the Office of Federal Procurement Policy (OFPP), is one such reform. It is applicable to some if not all aspects of the various types of projects operated by the department. Its effective use depends on the ability of the government/contractor team to understand and manage the risks inherent in any contracted effort.

PBC emphasizes that all aspects of an acquisition should be structured around the purpose of the work to be performed, as opposed to the manner in which it is performed. It offers contractors flexibility to determine how best to meet the government's requirements, while ensuring that desired performance levels are achieved and that payment is made only for results that meet negotiated performance standards. DOE has included this technique as part of its own contract-reform agenda and has been using a performance-based approach for major projects at Rocky Flats, Oak Ridge, the Nevada Test Site, and elsewhere.

Successful PBC is based on defining existing conditions, specific requirements, and the desired results or outcomes, along with objective, meaningful, and measurable performance and quality standards. In addition, incentives are used to focus contractor efforts and to reward success. In a successful performance-based contract, expectations must be made clear, with agency and contractor teams working together in a business partnership to achieve well-defined and measurable results.

Agencies across the government are increasingly relying on PBC. Moreover, a March 9, 2001, memorandum to all agencies from the deputy director of the Office of Management and Budget (OMB) requires that 20 percent by dollar value of all agency service contracts for FY2002 be performance-based. This mandated percentage would increase in future years.

A 1998 study conducted by OFPP found cost savings on the order of 15 percent and increases in customer satisfaction (up almost 20 percent) when agencies used performance-based contracts rather than traditional requirements-based procurements (EOP, 1998). Much of the success of a PBC approach results from effective use of a cross-functional team for identifying desired outcomes and establishing effective performance metrics.

Integrated Project Teams

The integrated project team (IPT) concept included in DOE Order O413.3 is an essential element in implementing a performance-based approach (DOE, 2000). The committee strongly supports the use of these IPTs and suggests the following PBC methodology:

- Establish a cross-functional government team of program, contracting, and project management personnel to work together to develop, award, implement, and manage a performance-based contract or task. If the contract has already been awarded, then a joint government/contractor team should develop useful performance metrics and incentives.
- Develop a performance-based statement of work (SOW), in matrix format, to be used as the basis for all subsequent solicitation, proposal, or contract documents.
- Define, for each major project or task, the desired outcomes, required services, standards of performance, and methods of performance evaluation and measurement.

While DOE has followed similar procedures in requiring performance metrics for service contracts, the General Accounting Office (GAO) has been critical of DOE's employment of PBC (GAO, 2001). It cited, among other things, inadequacies in DOE financial accounting and reporting systems and difficulties in establishing firm baselines from which to measure contractor performance. In its January 2001 report, the GAO continues to cite PBC implementation problems.

Balanced Scorecard

The DOE Office of Procurement and Assistance Management has conducted its own study of this area and has required headquarters review of the various performance metrics and incentives being used. In addition, a department-wide balanced scorecard self-assessment training program includes PBC. The balanced scorecard is an approach for measuring an organization's performance and long-term success. Measurements are made in the areas of finance, customer service, internal business processes, and employee learning and growth. This tool is in use in a number of agencies to measure performance and identify strengths and weaknesses.

For FY2000, 28 of DOE's major site and facility management contractors participated in a balanced scorecard process. Examples of performance metrics include a customer satisfaction index, with firms on average achieving a customer satisfaction score of 90 percent. BWX Technologies at the Mound site received the top score of 100 percent. From an internal business perspective, the performance measure used is the percentage of systems in compliance with stakeholder requirements. Here the average for DOE was 92 percent, with Bechtel Nevada achieving a 100 percent rating.

The scorecard has been in use for 4 years. All of the major site and facility contracts awarded in the past 4 years were performance based. For FY2001 a new hands-on training program was rolled out, with a new performance-based management-contracting course under development. This training should be included as part of the project manager training program discussed in Chapter 9, "Project Manager Training and Development."

Benefits of Performance-Based Contracting

PBC is not only a contracting technique, it is also critical to the requirements development process, because it entails defining desired outcomes up front as well as assessing risks. These steps cannot be taken without the full involvement of the IPT.

The committee believes that effective use of PBC will give DOE more confidence in the likely success of its major projects. To be fully effective, program managers need to be integrated into this process in every respect. The key benefits of adopting this approach are the following:

- Requirements and processes with no value added are eliminated, enabling the contractor to achieve objectives faster and at lower cost.
- Contractor innovation is encouraged.
- Expectations and accountability are clearly defined.
- A win-win partnership is established between the contractor and the customer, with risks and rewards shared.

PERFORMANCE-BASED CONTRACTING IN POLICY AND PROCEDURE DOCUMENTS

Given the potential impact of PBC on successful project management and the benefits to all players in this process, the committee believes that PBC objectives and methodology should be defined and discussed in DOE policy and procedure documents. The following subsections suggest the information that might be included in the draft PPM manual, along with examples of PBC applicability to the DOE environment.

Define Desired Outcomes

Outcome definition should be functional and substantive, using terms that can be easily understood by an external stakeholder and focused on results rather than on work processes. These include the desired outcomes for the overall contract, project, or task and the desired outcomes for major task areas.

Each of the major task-level outcomes should contribute to the overall program or project outcome. All outcomes should be based on business results for which the contractor can reasonably be held accountable and should not depend on performance or events outside the contractor's control.

Examples of outcomes for contractor purchasing systems that are currently included in the DOE balanced scorecard assessment are the following:

- On-time delivery is a measure of effective supplier management. The performance metric is the percentage of on-time delivery to be achieved, with the target set at 85 percent.

- Streamlining processes to enable DOE to more efficiently meet its mission needs. The performance metric is the number of critical processes reengineered, redesigned, or revalidated. In one case the target is two annually.

These examples demonstrate types of measures that may be used to see whether performance criteria have been achieved.

Define Required Services for Specific Milestones or Tasks

All required services should contribute to the achievement of the project or task-level outcome. These services might include major deliverables or work products. All requirements should focus on the outcomes to be provided and should not define the process or technical approach used to perform the work.

Define Performance Standards for Each Required Service

Each required service may have multiple associated performance standards. All performance standards should measure services for which the contractor can be held fully accountable and should be precise, meaningful, and attainable. The following considerations apply:

- Define performance standards only for those areas where the data collected are sufficiently meaningful to be evaluated and can be used to determine whether or not the desired results have been achieved. Consider the cost/benefit of collecting the performance data.
- Typical performance standards address quality, timeliness, completeness, accuracy, reliability, and cost.
- In determining the specific quantitative metrics, consider the cost/benefit of achieving varying levels of performance.
- Industry standards, departmental policies, and regulations that the government requires the contractor to comply with may be used as performance standards. This point is particularly relevant in the DOE context, since compliance with legal or regulatory directives is often a prerequisite for performing the work.

The performance metrics described above are examples of the ways in which success can be measured.

Define Acceptable Quality Levels for Each Performance Standard

Acceptable quality levels (AQLs) should be expressed as a percentage of conformance with the standard (e.g., 95 percent) or the deviation (e.g., 0 percent deviation), or as a range (e.g., not to exceed 2 days following a negotiated deadline). Every performance standard must have an AQL. The cost implications of requiring 100 percent compliance with or 0 percent deviation from the standard should be fully considered before setting tolerance levels. However, if safety or compliance requirements are at issue, then 100 percent compliance may be the only acceptable approach.

Define Surveillance or Monitoring Methods to Document Performance and Evaluate Whether Performance Standards and AQLs Have Been Met

There should be at least one surveillance or monitoring method for each performance standard. In general, surveillance includes both contractor-provided data and government validation. The government's primary role should be to validate and assess performance.

Identify Incentives and Disincentives Relating to the Critical Results

Disincentives, including loss of fee, can be used especially where activities relate to the health and safety of workers or the public or where serious environmental impacts may occur. While monetary incentives are generally used, other potential incentives may relate to reduced reporting requirements or to extending automatically the term of a contract based on a contractor's superior record of performance. The latter approach is called award term contracting. In the DOE context, with a relatively limited base of bidders, an award term approach may further limit competition and therefore may not be as suitable as it is at other agencies.

Frequently, in the DOE context, timeliness incentives are critical, whether they involve speeding cleanup operations at a site like Rocky Flats or an additional fee for accelerating a schedule, as is the case for the new contract at Yucca Mountain.

An innovative incentive approach currently in development by the National Nuclear Security Agency (NNSA) is the use of multisite integrated incentives for key programs. These multiyear incentives, which will become effective in FY2002, link the sites in achieving cost efficiencies.

As discussed above, effective PBC approaches have been developed through the balanced-scorecard self-assessment process. Other examples include the M&I contract at Oak Ridge, Tennessee, which uses effective performance-based tasks and subcontractor competition to effect its cleanup operations. Oak Ridge has also established a clear government-contractor partnership document to ensure accountability and to identify each party's responsibilities in getting the sought-after results. Similarly, the Rocky Flats, Colorado, cleanup operation has developed performance metrics and incentives that focus both the government and the contractor on the desired end state, that is, successful site cleanup and closure of operations. Finally, the Nevada Test Site contract has used specific, objective, fee-based performance incentives to focus all aspects of the contractor's operations on achieving the government's goals.

Develop a Performance Matrix to Assess PBC

As noted previously, the committee strongly supports the OECM policy to use IPTs to develop an acquisition plan for the contract and then to play a continuing role throughout the acquisition cycle. The IPT should develop a performance matrix that uses the categories of effort described above as column headings. This matrix will display desired business results, performance metrics, and performance monitoring or surveillance methods, as well as incentives or disincentives associated with the contract all in one easily readable and understandable document. The matrix in effect serves as the work statement for the contractor and should be incorporated into the resulting contract or task order. Any references, assumptions, or dependencies should also be included in the matrix to the extent possible.

As a practical matter, understanding the various dimensions of the PBC matrix can take some time. Moreover, it is essential that all team members have a common understanding of what each term means and how it is to be used. For this reason, the committee believes that PBC training should be provided to all IPTs, if possible using a just-in-time training method, so that all key players will be fully informed on the benefits of the approach as well as on its methods and objectives.

In sum, the performance matrix defines the performance and results expected by the government but does not dictate how the work is to be performed unless law or regulation mandates certain steps.

RISK, PERFORMANCE-BASED CONTRACTING,
AND CONTRACT TAILORING

A central part of the IPT's effort in developing the performance matrix is the need to consider risks very early in the acquisition strategy and acquisition planning process. Only after the risks have been fully assessed can the contracting approach and type of contract be determined. PBC supports fixed-price contract vehicles by focusing on results as opposed to level of effort. The contractor is given considerable operational leeway but is held strictly accountable for meeting the outcomes established by the IPT for tasks and for the overall effort. Of course, by moving toward a fixed-price environment, the government is effectively transferring risk to the contractor.

As the level of uncertainty surrounding technical risk (e.g., uncharacterized elements in a cleanup operation or scalability and integration issues in an information technology or scientific environment) or business risk (e.g., the likelihood of adequate funding to achieve project milestones) increases, the manageability of the risk comes into question. Where the uncertainties are too great, the contractor is less likely to be held accountable for the desired result. The same is true if interdependencies cannot be adequately identified.

In a fixed-price environment, the government transfers certain cost risks to the contractor, as defined by the contract specification. That is, the fixed price puts a ceiling on the cost to the government for risks deemed to fall under the contractor's control or risks considered normal in the particular line of business (weather and other things), provided the risks are identified in the contract. However, a fixed-price contract is also a floor on the cost to the government, unless the project is adequately managed to eliminate changed conditions, delays, and other sources of additional cost to the owner. Also, the risk of delays, schedule overruns, or quality of the facility may devolve to the government unless the required performance and desired outcomes are clearly defined and monetary or other incentives provided for superior performance. Risks may also revert to the government if there are changed conditions or design changes for which the government may be held responsible. Again, the PBC matrix provides a very effective tool for assuring contractor accountability for results.

In an environment where uncertainty is great, fixed-price contracts would be unreasonable or imprudent. That docs not mean, however, that the contracted effort could not be structured in such a way as to allow performance-based approaches and competition for those aspects of the effort where risk is both clear and manageable.

DOE is undertaking an acquisition risk management (ARM) study to be completed by December 2001 that focuses on the points included in DOE O413.3 (DOE, 2000) and Part 7 of the Federal Acquisition Regulation (FAR) that project risks need to be managed early in the acquisition planning stage. The study will result in a guide to be used by IPTs in the development of a project's acquisition strategy. This guide should also help the team in deciding the amount of contingency to be included in execution estimates.

As an IPT works through these questions and issues, it will have much greater confidence that it has in fact structured an appropriate contracting approach for meeting DOE's mission. This front-end planning effort requires both time and the right mix of staff to see that all key questions are identified and addressed, for if these questions are not addressed early in the project, the prospects for problems later on are much greater.

While incentive contracting has been identified as a key tool for achieving cost savings and better results, the IPT should also consider other contracting methods to see if they might be more effective in meeting the department's needs. For example, at Los Alamos National Laboratory (LANL), the M&O contractor used a straightforward design/build project delivery method for the Advanced Computing Facility and more recently for the Non-Proliferation Center Building. Where the work to be done can be defined in traditional design and construction terms, as they were here, schedule advantages, project quality, and cost savings can be realized.

More information on design/build and other contracting modes that might meet DOE contracting needs are found at Part 36 of the FAR, "Construction and Architect-Engineer Contracts." In summary, defining desired outcomes and performance standards, as well as assessing risk early on in the acquisition process, is key to selection of the right contract vehicle and offers the best chance of meeting agency and project and program officer expectations. DOE should consider these innovative contracting methods when looking for the best method to achieve the outcomes sought.

A recurring issue in performance-based contracting is how to set a price on performance. In general, this might be done in one of two ways:

- The owner is highly knowledgeable, has an excellent database, and therefore knows the appropriate costs for a given level of performance and can negotiate equitable prices with prospective contractors.
- The owner is not knowledgeable and relies on the marketplace (that is, competitive bidding) to set the price.

Clearly, the second method requires a large pool of responsible bidders. The Phase II report noted that the DOE bidder pool was shrinking and recommended that DOE act to arrest and reverse this trend. In 2001, a report by the DOE Contract Reform and Privatization Project Office, *Analysis of the DOE Contractor Base: Readiness, Willingness, Profitability, and Trends: A Focus on the Environmental Management Program*, showed that for EM, at least, the contractor base has shrunk even further since 1999 (DOE, 2001). To pursue performance-based contracting effectively, DOE should either (1) take steps to increase the contractor base in order to carry out the second method or (2) fall back on the first method and become an owner knowledgeable about costs and performance (or both). Execution of the first method is very difficult if only the contractors know the costs.

FINDINGS AND RECOMMENDATIONS

Finding. The extent of training and use of PBC in DOE contracting efforts is unclear. There is no DOE-wide database that shows the extent of use of PBC or the number of staff trained in PBC techniques.

Recommendation. The committee reaffirms the recommendations made in previous reports (NRC 1999, 2001) on using PBC and encourages OECM to play a lead role in supporting this practice. OECM should work closely with the Office of Procurement and Assistance Management to see that PBC training is provided as part of the career development process for project management personnel and just-in-time training for the IPT. In the near term, OECM should bring on board a cadre of experts, skilled in performance-based contracting, to provide technical assistance to IPTs responsible for new major system initiatives.

Finding. The draft *Program and Project Management* (PPM) manual and draft *Project Management Practices* (PMP) developed by the OECM fail to address PBC adequately.

Recommendation. The detailed descriptions of PBC alternatives and their application to DOE projects should be included in the revised PMP and PPM.

Finding: There have been continuing efforts on the part of DOE to move toward a more effective use of PBC methods and to support these efforts.

Recommendation. Contract approaches should be tailored to use fixed-price and performance-based methods where practicable to assist the DOE to get the most cost-effective results and to stimulate competition. In addition, the department should continue to explore other innovative commercial contracting approaches to meet its needs.

REFERENCES

DOE (Department of Energy). 2000. Program and Project Management for the Acquisition of Capital Assets (Order O413.3). Washington, D.C.: Department of Energy.

DOE, Contract Reform and Privatization Project Office. 2001. Analysis of the DOE Contractor Base: Readiness, Willingness, Profitability, and Trends: A Focus on the Environmental Management Program. Washington D.C.: Department of Energy.

EOP (Executive Office of the President). 1998. A Report on the Performance-Based Service Contracting Pilot Project. Washington, D.C.: Executive Office of the President.

GAO (General Accounting Office). 2001. Major Management Challenges and Program Risks: Department of Energy. GAO/GAO-01-246. Washington, D.C.: Government Printing Office.

NRC (National Research Council). 1998. Assessing the Need for Independent Project Reviews in the Department of Energy. Washington, D.C.: National Academy Press.

NRC. 1999. Improving Project Management in the Department of Energy. Washington, D.C.: National Academy Press.

NRC. 2001. Improved Project Management in the Department of Energy. Letter report, January. Washington, D.C.: National Academy Press.

8

Documentation of Project Management Policies and Procedures

INTRODUCTION

The Phase II report (NRC, 1999) noted that DOE did not have adequate policies and procedures for planning and managing projects and that no single authority was responsible for ensuring that project management policies and procedures were followed. It recommended that as a part of its project management system, DOE should issue fundamental policies, procedures, models, tools, techniques, and standards.

The committee noted in its January 2001 letter report that DOE had begun development of more effective project management policies, procedures, models, tools, techniques, and standards (NRC, 2001). In particular, DOE O413.3 has been issued, and drafts of the *Program and Project Management* (PPM) manual and *Project Management Practices* (PMP) have been reviewed by the committee (DOE, 2000a, 2000b, 2000c). OECM has advised the committee that because the documents were developed concurrently, there are discontinuities, overlaps, and repetitions that will be resolved in the next revisions. OECM reported that in the next revisions, the order will define policy, the manual will specify procedures, and the practices will provide commentary and examples. OECM is considering publishing the practices as a Web-based information system to be updated frequently. Although the documents are in the process of being revised, the committee believes there is value in providing the following comments and recommendations on the drafts.

OECM is to be commended on the significant progress in getting P413.1 and O413.3 produced and published. In the final analysis, the effectiveness of these or any policies and procedures depend on a commitment by DOE leadership to the task of continuous improvement of DOE project performance using the proven project management methods and techniques set out in these documents.

DOE POLICY P413.1 AND ORDER O413.3

Order O413.3 provides direction for program and project management for the acquisition of capital assets within the department. The order supplements P413.1, which established OECM, and provides additional detail on the office's roles and responsibilities to support the deputy secretary as the secretarial acquisition executive in oversight of capital asset acquisition (DOE, 2000a, 2000d).

As a part of the process of implementing Order O413.3, OECM has conducted a number of meetings and workshops to respond to questions and comments and to obtain recommendations for improvements to O413.3 and related documents. The committee has reviewed O413.3 and conducted meetings at which DOE personnel and contractors gave presentations on the implementation of O413.3. The committee observes that O413.3 has proven effective in defining and implementing a number of fundamental and beneficial changes for the department that will improve long-term project performance; however, there are several clarifications, improvements, and adjustments that, while not changing the basic policy, would improve it. The committee therefore offers the following observations:

DOE should define the term "baseline," as used in O413.3, to reflect the exact use of this term by Congress and the departmental obligations to manage projects to baseline values for scope, cost, and schedule.

- The relationships among the department's procurement, acquisition, and project management processes need clarification. The integrated project team (IPT) provides a framework for both acquisition and project management processes, but the policies and procedures on the IPT organization and requirements should be clarified and strengthened.
- O413.3 should make it clear that there are no exceptions to the requirement that each project have a thorough, documented project plan to support every step in its approval process.
- As recommended in the Phase II report, to increase the uniformity of information across the complex, the policy should require that all projects apply an earned value management system (EVMS) for reporting project performance. The policy should define the department's EVMS reporting requirements and the cost accounting systems necessary to support EVMS so that they can be applied to all types of projects. Only short projects with total durations of less than 6 months should be excluded from this requirement.

DRAFT *PROGRAM AND PROJECT MANAGEMENT* MANUAL AND DRAFT *PROJECT MANAGEMENT PROCEDURES*

Purpose

The purpose of the PPM and PMP should be to describe what is needed to pass through the DOE critical decision planning gates. The committee believes that the documents should add value by defining methods to streamline procedures and improve project performance. The PPM should be prescriptive, while the PMP should provide guidance. The committee observed that the roles of the two draft documents were intermixed. While the documents can provide general guidance, they need to be specific where compliance is intended.

Format and Content

Overall, the two documents should be formatted to allow easy cross-reference to O413.3 and from one document to the other. There should be a logical organization that would allow project managers to progress through the two documents referring from one to the other and at the conclusion to know what the department requires and what tools to use.

The draft PPM provides an overview of the DOE project processes, including the deliverables at decision points and management responsibility and authority, but the detailed information is not organized according to a project delivery process. OECM has indicated that reorganization according to the DOE process as defined in O413.3 is planned. PPM Chapters 7 through 17 appear to follow the Project Management Institute's (PMI's) *Project Management Body of Knowledge* (PMBOK) (PMI, 2000). The committee believes that generic project management information should not be in the documents. The source of this information could be referenced or the PMBOK could be provided along with the DOE documents. These chapters include valuable ideas but these ideas need to be integrated into the DOE processes rather than being listed separately. The same problems are evident in the draft PMP.

The draft documents were overwhelming in their detail in some sections; in some instances the information was superfluous and in others misleading. Conversely, some important issues received very little attention in the PPM—for example, the 70 percent of design engineering that is to be accomplished after the preliminary design. Managing the design phase of engineering and controlling changes to the initial performance baseline are of critical importance to the final outcome of the project. The discussion of DOE change control procedures is also inadequate.

Some important issues are missing entirely from the draft PMP and PPM. These include team alignment and teamwork procedures for including stakeholder and public participation; project scope definition; and control of scope change. The critical decision information packages in the PMP should be described instead in the PPM, because the critical decision points and documentation packages required by O413.3 are not standard project management practices.

The PMP should focus on detailed practices in a manner that assures they are given attention commensurate with their respective significance. Because the PMP is largely generic, common knowledge (much of it replicates the PMBOK, for example), the space it gives to a topic is often inversely proportional to its relevance to DOE. For example, 87 pages are spent discussing value engineering, an important but generally well-understood practice, and only 8 pages are given to planning the project's acquisition strategy, an issue that is fundamental to the success of a project.

Revisions to the draft PMP document should center on the characteristics of successful project performance, how it is achieved, and what constitutes a best project management practice for DOE (see Appendix C in the Phase II report (NRC, 1999)). DOE best practices should be documented to provide a how-to resource rather than a didactic educational resource. Tailoring each practice to DOE's missions and processes is also very important. Achieving predictable, consistent, repeatable project performance results through standards and prescribed practices is essential.

Much of the material presented as general practices represents the authors' preference. Different authors would have different opinions. It should be made clear which material constitutes DOE policies, procedures, or preferences and which is merely personal advice from the authors.

The PPM should describe what outcomes and deliverables DOE management expects. It should establish basic reporting standards, documentation content, use of M&O and M&I contractor services, and organizational structures. It needs to establish clear roles and responsibilities for project and operational managers. The process details should be prescribed with three things in mind: (1) every project will not always have an experienced federal project manager (FPM), (2) not all projects require the same degree of detail in the reports, and (3) simplicity, particularly with respect to communication, is crucial to success.

ISSUES IN NEED OF ADDITIONAL DOCUMENTATION

Some specific issues that were identified in the Phase II report and which the committee considers to be particularly in need of more focused attention in the policies and procedures documentation are discussed individually below.

Value Engineering

Value engineering (VE) is required for federal agencies, and a DOE-wide VE program was recommended in the Phase II report and reaffirmed in the committee's 2001 letter report (NRC 1999, 2001). DOE O414.3 requires VE, but the committee has yet to see any statistics indicating that the use of VE on DOE projects has increased beyond the inconsequential number reported in 1999. The draft PPM and PMP address VE, but the goal of a department-wide organization to support the consistent application of VE for all projects has not been addressed. In general, VE is a well-understood process, but the DOE procedure documents do not specify who should do it, when it should be done, whether it should be a discrete step or a continuous process, whether it should be done once or repeated if there have been design or scope changes, and whether it should be a separate effort or combined with EIRs and/or ICRs. These questions, and others, should be answered in the documentation, and the procedures should clearly require that VE reports be made a part of the project's baseline review and critical decision process.

Change Management

The Phase II report included the recommendation that DOE establish a system for managing change that provides traceability and visibility for all baseline changes (NRC, 1999). DOE O413.3 defines levels of authority but defers definition of the thresholds for change control to the project execution plan (PEP) (DOE, 2000a). The PPM states as follows: "Project changes shall be identified, controlled, and managed through a traceable, documented, and dedicated change-control process," but it also defers definition of the process to the PEP (DOE, 2000b). The committee has observed several instances of baseline change without the documentation of a traceable and verifiable process; however, the projects were initiated before the current policies and procedures were issued. The project change control process is critical to on-time and on-budget performance and remains an ongoing concern of the committee. The committee believes that the change control process should be better defined in the DOE policies and procedures to facilitate change control on all projects.

ISO 9000 CERTIFICATION

The International Organization for Standardization's (ISO) 9000 certification is, in essence, a quality assurance function that, in part, compares the actual processes used by an organization with the organization's documented policies and procedures. The latest version of ISO 9000, adopted in 2000, goes beyond a comparison of documentation and practices to require a continuous process improvement program, as discussed in Chapter 2 and Chapter 5 (ISO, 2000). The Phase II report recommended that DOE offices should obtain and maintain ISO 9000 certification of all its project management activities (NRC, 1999). This recommendation was reaffirmed in the January 2001 letter report (NRC, 2001).

The committee recognizes that the PMSOs, especially in DP, and OECM have taken steps to align the actual processes with published procedures. Nevertheless, it is highly unlikely that DOE could pass an ISO 9000 audit today. DOE O413.3 has been released, but the implementing PPM and PMP are still in the draft stage and many months overdue. The committee is reliably informed that many DOE project management personnel continue to refer to DOE Order 4700.1, *Project Management System*; this document may provide excellent advice, but it was superseded in 1995 by DOE Order 430.1, *Life Cycle Asset Management*, itself now superseded. In short, it is clear that many DOE personnel are engaged in processes that have no officially recognized and documented procedures. It is difficult to see how anyone can be accountable for following processes that are nowhere documented.

The committee continues to believe that the process involved in preparing for and obtaining ISO 9000 certification would have direct benefits for the department, and it recommends that the deputy secretary give serious consideration to making ISO 9000 certification a DOE goal.

FINDINGS AND RECOMMENDATIONS

Finding. The recommendations in the Phase II report to develop and publish a set of policies and procedures for management of DOE projects appear to have been addressed to some extent by the draft PPM and the draft PMP; however, the committee finds that there is a need for additional detail and clarity and elimination of discontinuities, gaps, overlaps, and repetitions. The committee recognizes that OECM is addressing these issues as it develops the next iterations and commends OECM for its leadership role.

Recommendation. The PPM and PMP text should be tailored to specific DOE requirements. It should be clear which parts of the text constitute DOE required procedures and which parts reflect general advice on good project management practices.

Recommendation. OECM should assure that policies and required procedures add value by streamlining the process and improving project performance. Policies and procedures that do not demonstrably add value should be revised or eliminated.

Recommendation. The PPM and PMP should have parallel structures. A complete index and a glossary of terms should be provided for both documents.

Recommendation. Examples should be given where they will illustrate the application of procedures and the necessary documentation. Examples should have adequate explanations and represent realistic project situations. Over time, a set of templates and case studies should be built up.

Recommendation. OECM should be provided the resources needed to publish improved, revised versions of the PMP and PPM as soon as possible. OECM should be given the authority to authorize case-by-case exceptions when appropriate to ensure that common sense and cost-effectiveness prevail in the retrofitting of procedures to ongoing projects.

REFERENCES

DOE (Department of Energy). 2000a. Program and Project Management for the Acquisition of Capital Assets (Order O413.3). Washington, D.C.: Department of Energy.

DOE. 2000b. Program and Project Management. Draft. Washington, D.C.: Department of Energy.

DOE. 2000c. Project Management Practices. Draft. Washington, D.C.: Department of Energy.

DOE. 2000d. Program and Project Management Policy for the Planning, Programming, Budgeting, and Acquisition of Capital Assets (Policy P413.1). Washington, D.C.: Department of Energy.

ISO (International Organization for Standardization). 2000. ISO 9000:2000. Quality Management Systems—Fundamentals and Vocabulary. Geneva, Switzerland: International Organization for Standardization.

NRC (National Research Council). 1999. Improving Project Management in the Department of Energy. Washington, D.C.: National Academy Press.

NRC. 2001. Improved Project Management in the Department of Energy. Letter report, January. Washington, D.C.: National Academy Press.

PMI (Project Management Institute). 2000. Project Management Body of Knowledge (PMBOK). Newton Square, Penn.: Project Management Institute.

9

Project Manager Training and Development

INTRODUCTION

The Phase II report found that the competencies needed for successful project managers were lacking in the DOE, and that this was a fundamental cause of poor project performance. This situation largely emanates from the absence of a career program and the lack of training and development opportunities for project management professionals. The Phase II report recommended the establishment of a department-wide training program, as well as criteria and standards for selection and assignment of project managers, including requirements for training and certification (NRC, 1999). In earlier chapters of this report the committee identified the urgent need for specific training in front-end planning, risk management, EVMS, and performance-based contracting.

It is reported that a lack of confidence in currently available training programs, limited staffing, and the decision to concentrate efforts on other project management deficiencies have delayed action on improving training and development. The committee believes that as a consequence, progress on enhancing the competencies of project managers has been inadequate. A task force for project management career development, chartered to address the Phase II report recommendations, has done a commendable job; however, much remains to be done. A commitment from top management and additional resources will be needed to implement the training and development programs being planned by the task force. The training, development, and retention of qualified project managers will continue to be a major challenge for DOE.

PROJECT MANAGEMENT CAREER DEVELOPMENT PROGRAM

In January 2001, the deputy secretary of DOE directed OECM to lead a 2-year effort to develop and implement the project management career development program (PMCDP). To accomplish this goal, a task force was established that includes representatives from PSO headquarters and field offices and experts from other federal agencies. The task force is supported by OECM personnel, contractors, and rotating personnel from other DOE offices. It has gathered, analyzed, and synthesized much of the information needed to create a training program and readied some tasks for implementation. The committee applauds the task force effort to create a program geared to developing the knowledge and skills needed by project managers to fulfill the missions of the agency. Significant accomplishments of the task force to date include the following:

- An inventory of project managers;
- A benchmarking study of best practices for project management career development in other federal agencies and industry;
- Documentation of the roles and responsibilities of DOE project managers;
- A partial matrix diagram of the knowledge and skills required for 5 competency levels in 10 domains (general project management, leadership/team building, scope management, communication management, quality/safety management, cost management, time management, risk management, contract management, integration management);
- Identification of training and experience requirements for each of the 10 domains (in progress);
- A gap analysis of current levels of experience, education, and skills (in progress); and
- Descriptions of training courses for each of the 10 domains (in progress).

Some significant tasks remain to be accomplished:

- Complete the matrix diagram of knowledge and skills.
- Complete the gap analysis.
- Contract for training course development/delivery.
- Develop the training curriculum.
- Develop experience histories of project managers.
- Conclude the effort on implementing a certification requirement.
- Integrate the tracking of competencies, certification, and training into the Corporate Human Resources Information System.

INTERIM TRAINING EFFORTS

The committee is concerned that it will take fully 4 years from the time the Phase II report was issued for activities to begin that address the critical need for project manager career development. It appears that the process of developing a program has been accepted as the solution to the problem. The committee believes that there should be active training while the plan to undertake a refined program is developed.

Even though the final curriculum for project manager training will not be completed until next year, it is imperative that training not be neglected in the interim. The committee urges that training be escalated, particularly in those areas and for those individuals where known shortfalls exist. The committee acknowledges that the PSOs recognize the need for training and are implementing it to various degrees and that existing courses are being revised to meet present needs. However, it is important that these activities be given higher priority in response to the deficiencies revealed by the DOE gap analysis and issues identified by this report: specifically, in front-end planning, risk management, and performance-based contracting. The committee believes that project management training expenditures should be increased to a level comparable to that reported by the American Society for Training and Development (ASTD) for similar training in the private sector (ASTD, 2001). Management should ensure that resources are available and that participation in training programs is mandatory.

ALTERNATIVE LEARNING CONCEPTS

The committee considers the extant department-wide training contract, which gives exclusive rights for providing training, to be an impediment to obtaining training that is timely and available in various formats and alternative learning concepts. In the long term, sole-source contracts of this nature cannot meet DOE's needs. Training courses for DOE project managers should be taught by personnel with extensive experience in managing projects.

The development and deployment of alternative learning concepts are needed to impart accessible, timely, high-quality information to project managers. There is a need for flexible approaches that can fit project managers' locations, schedules, and levels of experience. Several alternatives to traditional classroom learning have been developed and used successfully for professional development (Dixon, 1998):

- *E-learning.* E-learning refers generically to the use of CD-ROMs, computer-based learning, and various forms of Web-based learning. Many universities now offer courses and degree programs via the Internet.

- *Action learning.* Action learning describes a program whereby groups of colleagues (learning sets) are brought together in real time by electronically mediated means to work on real workplace problems. Action learning is a systematic approach to learning while solving real problems at work. While action learning is individually focused, it uses a small group, known as a learning set, which provides a forum where each set member's ideas can be discussed and challenged in a supportive environment. Action learning is an iterative, experiential process, involving a cyclical notion of learning. The elements of an action learning cycle are the following (McGill and Beaty, 1992):
  —An action;
  —Reflection—consideration of the effects, successful and unsuccessful, of that action;
  —Generalizing—identifying new learning from this experience that can be applied; and
  —Planning—deciding on the basis of generalizations how to act in the future.
- *Just-in-time learning.* The application of just-in-time learning for project teams is a means to deliver relevant information and improve team coordination. DP is planning to activate a just-in-time training program using a system developed in private industry.
- *Learning portfolios.* DOE project managers should be encouraged to develop learning portfolios. Learning portfolios are portfolios containing evidence of learning, work experiences, and achievements, for a specific learning goal. Learning goals can be established by an employee with the assistance of a mentor or supervisor. Portfolios may include a variety of documents, such as descriptions of projects, personal audits, research papers or articles, diaries of relevant experiences, notes from consultations with colleagues, case descriptions, and certificates from formal training programs.

OTHER CONCERNS

The PMCDP is directed at the training and development of project engineers; however, the committee believes that enhancing abilities in project management should also be directed at a broader audience of project management-related personnel, including program managers, support personnel, upper management, and contract project managers. Another issue is that federal agencies, including DOE, are facing a crisis brought on by the aging and impending retirement of experienced personnel. The aging workforce makes it even more urgent to develop a new group of competent project managers from younger, less experienced personnel and to create a career development program that will help retain skilled project managers.

FINDINGS AND RECOMMENDATIONS

Finding. Although there is a clear and immediate need to provide project management training, courses developed under the current PMCDP effort will not be available until late 2003. Training is the equivalent of providing workers the tools to accomplish their job.

Recommendation. DOE should immediately implement an accelerated training program to improve the knowledge, skills, and abilities of project managers to address recognized gaps while continuing the PMCDP planning effort. Immediate measures should be taken to eliminate impediments and use current resources to explore creative and cost-effective nonclassroom alternatives such as e-learning, action learning, and learning portfolios. Also, trainers skilled in specific topics should be engaged to instruct a cadre of DOE employees, who in turn will impart department-wide training to other DOE employees.

Recommendation. At the beginning of each fiscal year, DOE management should budget the funds to accomplish the projected training objectives for that year and should persist in mandating the accomplishment of individual career development objectives.

Finding. The existing contract for training offers a means to deliver consistent content throughout the department; however, it reduces the range of options for training.

Recommendation. DOE should modify or replace the current contract to allow greater flexibility in accessing courses pertinent to the project management skills utilized by industry and other federal agencies. DOE should develop new courses consistent with the new knowledge, skills, and abilities requirements identified by the findings of the gap analysis.

REFERENCES

ASTD (American Society for Training and Development). 2001. ASTD Benchmarking Forum. Available online at <http://www.astd.org/virtual_community/research/bench/benchmarking_forum_main.html>.

Dixon, R.L. 1998. "Action Learning: More Than Just a Task Force." *Performance Improvement Quarterly* 11(1): 44-48.

McGill, I., and L. Beaty. 1992. Action Learning : A Practitioner's Guide. London, U.K.: Kogan.

NRC (National Research Council). 1999. Improving Project Management in the Department of Energy. Washington, D.C.: National Academy Press.

Appendixes

A

Biographies of Committee Members

Kenneth Reinschmidt (National Academy of Engineering) is A.P. and Florence Wiley Professor of Civil Engineering at Texas A&M University and retired from Stone & Webster as senior vice president. He was appointed chair of this committee for his combination of expertise in the disciplines of civil engineering, project management, cost estimating, and the management of large-scale construction projects, including nuclear and fossil fuel power plant construction. He held various positions at Stone & Webster, including president and CEO of Stone & Webster Advanced Systems Development Services, Inc., and manager of the consulting group in the Engineering Department. In these positions he was engaged in structural engineering, operations research, cost analysis, construction engineering and management, and project management. Prior to his work at Stone & Webster, Dr. Reinschmidt was a senior research associate and associate professor in the Civil Engineering Department at the Massachusetts Institute of Technology, where he was engaged in interdisciplinary research on power plant engineering, design, construction, and project management. Dr. Reinschmidt served as chair of the committee that produced the recent NRC report *Improving Project Management in the Department of Energy* and was reviewer of the NRC report *Assessing the Need for Independent Project Reviews in the Department of Energy*. He is a former member of the Building Research Board of the NRC and served on or chaired several other NRC committees, including the Committee on Integrated Database Development, the Panel for Building Technology, the Committee on Advanced Technology for Building Design, and the Committee on Foam Plastic Structures. He has also served on several National Science Foundation review panels on construction automation, computer-integrated construction, and engineering research centers. He obtained his B.S., M.S., and Ph.D. degrees from the Massachusetts Institute of Technology.

Don Jeffrey (Jeff) Bostock recently retired from Lockheed Martin Energy Systems, Inc., as vice president for engineering and construction with responsibility for all engineering activities in the Oak Ridge nuclear complex. He served on this committee because of his experience with managing projects as a DOE contractor. He has also served as vice-president of defense and manufacturing and manager of the Oak Ridge Y-12 plant, a nuclear weapons fabrication and manufacturing facility. His career at Y-12 included engineering and managerial positions in all of the various manufacturing, assembly security, and program management organizations. He also served as manager of the Paducah Gaseous Diffusion Plant, which provides uranium enrichment services. He was a member of the committees that produced the NRC reports *Proliferation Concerns: Assessing U.S. Efforts to Help Contain Nuclear and Other Dangerous Materials and Technologies in the Former Soviet Union* and *Protecting Nuclear Weapons Material in Russia*. Mr. Bostock also served as a panel member for the annual NRC assessment of the Measurement and Standards Laboratories of the National Institute of Standards and Technology. Mr. Bostock has a B.S. in industrial engineering from Pennsylvania State University and an M.S. in industrial management from the University of Tennessee. He is a graduate of the Pittsburgh Management Program for Executives.

Donald A. Brand (National Academy of Engineering) retired from the Pacific Gas and Electric (PG&E) Company as senior vice president and general manager, Engineering and Construction Business Unit, and is currently a lecturer in the Department of Civil Engineering at the University of California, Berkeley. Mr. Brand was appointed as a member of this committee because of his expertise in the management of the design, engineering, and construction of large, complex energy-related facilities. During his 33 years with PG&E, he carried out numerous managerial and engineering responsibilities related to the design, construction, and operation of fossil fuel, geothermal, nuclear, and hydroelectric generating facilities, as well as to electrical transmission, distribution, and power control facilities. Mr. Brand's industry activities have included membership on the Electric Power Research Institute's Research Advisory Committee and on the Association of Edison Illuminating Companies' Power Generation Committee. He received a B.S. in mechanical engineering and an M.S. in mechanical (nuclear) engineering from Stanford University. He also graduated from the Advanced Management Program of the Harvard University School of Business.

Allan V. Burman is president of Jefferson Solutions, a division of the Jefferson Consulting Group, a firm that provides change management services and acquisition reform training to many federal departments and agencies. He served as a member of this committee because of his expertise in federal acquisition, procurement, and budget reform. Dr. Burman provides strategic consulting services to private sector firms doing business with the federal government as well as to federal agencies and other government entities. He also has advised firms, congressional committees, and federal and state agencies on a variety of management and acquisition reform matters. Prior to joining the Jefferson Consulting Group, Dr. Burman had a lengthy career in the federal government, including serving as administrator for federal procurement policy in the Office of Management and Budget (OMB), where he testified before Congress over 40 times on management, acquisition, and budget matters. Dr. Burman also authored the 1991 policy letter that established performance-based contracting and greater reliance, where appropriate, on fixed-price contracting, as the favored approach for contract reform. As a member of the Senior Executive Service, Dr. Burman served as chief of the Air Force Branch in OMB's National Security Division and was the first OMB branch chief to receive a Presidential Rank Award. Dr. Burman is a fellow and member of the board of advisors of the National Contract Management Association, a principal of the Council for Excellence in Government, a director of the Procurement Round Table, and an honorary member of the National Defense Industrial Association. He is also a contributing editor and writer for Government Executive magazine. Dr. Burman obtained a B.A. from Wesleyan University, was a Fulbright Scholar at the Institute of Political Studies, University of Bordeaux, France, has a graduate degree from Harvard University and a Ph.D. from the George Washington University.

Lloyd A. Duscha (National Academy of Engineering) retired from the U.S. Army Corps of Engineers in 1990 as the highest-ranking civilian after serving as deputy director, Engineering and Construction Directorate, at headquarters. He served as a member of this committee because of his expertise in engineering and construction management and his roles as principal investigator for the NRC report *Assessing the Need for Independent Project Reviews in the Department of Energy* and a member of the committee that produced the NRC report *Improving Project Management in the Department of Energy*. He served in numerous progressive Army Corps of Engineer positions in various locations over 4 decades. Mr. Duscha is currently an engineering consultant to various national and foreign government agencies, the World Bank, and private sector clients. He has served on numerous NRC committees and recently served on the Committee on the Outsourcing of the Management of Planning, Design, and Construction Related Services as well as the Committee on Shore Installation Readiness and Management. He now chairs the NRC Committee on Research Needs for Transuranic and Mixed Waste at Department of Energy Sites. He has also served on the Board on Infrastructure and the Constructed Environment and was vice chairman for the U.S. National Committee on Tunneling Technology. Other positions held were president, U.S. Committee on Large Dams; chair, Committee on Dam Safety,

International Commission on Large Dams; executive committee, Construction Industry Institute; and the board of directors, Research and Management Foundation of the American Consulting Engineers Council. He has numerous professional affiliations including a fellowship in the American Society of Civil Engineers and in the Society of American Military Engineers. He holds a B.S. degree in civil engineering from the University of Minnesota, which awarded him the Board of Regents Outstanding Achievement Award.

G. Brian Estes is the former director of construction projects at the Westinghouse Hanford Company, where he directed project management functions supporting operations and environmental cleanup of the Department of Energy Hanford nuclear complex. He was appointed as a member of this committee because of his experience with DOE, as well as other large-scale government construction and environmental restoration projects. He served on the committee that produced the recent NRC report *Improving Project Management in the Department of Energy* and has served on a number of other NRC committees. Prior to joining Westinghouse, he completed 30 years in the Navy Civil Engineer Corps, achieving the rank of rear admiral. Admiral Estes served as commander of the Pacific Division of the Naval Facilities Engineering Command and as commander of the Third Naval Construction Brigade at Pearl Harbor. He supervised over 700 engineers, 8,000 Seabees, and 4,000 other employees in providing public works management, environmental support, family housing support, and facility planning, design and construction services. As vice commander, Naval Facilities Engineering Command, Admiral Estes led the total quality management transformation at headquarters and two updates of the corporate strategic plan. He directed execution of the $2 billion military construction program and the $3 billion facilities management program while serving as deputy commander for facilities acquisition and deputy commander for public works, Naval Facilities Engineering Command. He holds a B.S. in civil engineering from the University of Maine and an M.S. in civil engineering from the University of Illinois and is a registered professional engineer in Illinois and Virginia.

David N. Ford is an assistant professor of civil engineering at Texas A&M University. He served as a member of this committee because of his expertise in evaluating project management with analytical methods and simulations. He researches the dynamics of project management and the strategy of construction organizations, as well as teaching project management and computer simulation courses. Current research projects include an investigation into the causes of failures to implement fast-track processes and the value of contingent decisions in project strategies. Prior to his appointment at Texas A&M, Dr. Ford was an associate professor in the Department of Information Sciences at the University of Bergen in Norway. He was one of two professors to develop and lead the graduate program in the system dynamics methodology for 4 years. Dr. Ford's research during this time focused on the dynamics of product development processes and included work with Ericsson Microwave to improve that company's product development processes. Dr. Ford designed and managed the development and construction of facilities during 14 years in professional practice for owners, design professionals, and builders. The projects varied in size and facility type, including commercial buildings, residential development, industrial, commercial, and defense facilities. He serves as a reviewer for the journals *Management Science*, *Journal of Operational Research Society*, *Technology Studies*, and *System Dynamics Review*. Dr. Ford received his B.C.E. and M.E. degrees from Tulane University and his Ph.D. from the Massachusetts Institute of Technology in dynamic engineering systems.

G. Edward Gibson is an associate professor of civil engineering, associate chairman for architectural engineering, and the Fluor Centennial Teaching Fellow in the Construction Engineering and Project Management program at the University of Texas at Austin. He served as a member of this committee because of his expertise and research in preproject planning, organizational change, and the development of continuing education training programs for project managers. His research interests include organizational change, preproject planning, construction productivity, electronic data management, and automation and robotics. Dr. Gibson heads up the owner/contractor work structure thrust area of the Center for Construction Industry Studies funded by the Alfred P. Sloan Foundation. He received the Outstanding Researcher Award of the Construction Industry Institute (CII) for his pioneering work in preproject planning and is an author or coauthor of numerous articles and reports on this subject, including the CII *Pre-Project Planning Handbook* and the CII *Project Definition Rating Index* (PDRI). He also developed several CII education modules for continuing education and has taught over 125 short courses to industry in such areas as objective setting, team alignment, continuous improvement, preproject planning, and materials management. He received an M.B.A. from the University of Dallas and a B.C.E. and a Ph.D. in civil engineering from Auburn University.

Paul H. Gilbert (National Academy of Engineering) is senior vice president, principal professional associate, and principal project manager of Parsons Brinckerhoff Quade & Douglas, Inc., and director and senior vice president of Parsons Brinckerhoff International, Inc. He recently retired as director of Parsons Brinckerhoff, Inc., and chairman of Parsons Brinckerhoff Quade & Douglas, Inc. He served as a member of this committee because of his expertise in project management of design and construction of DOE facilities. Mr. Gilbert was the project director of the PB/MK Team for design, construction management, and construction of the conventional facilities of the Department of Energy's Superconducting Super Collider. He has served as principal in charge for major civil engineering projects such as the Stanford Linear Accelerator Positron-Electron

Project; the Basalt Waste Isolation Project at Hanford; the Nuclear Power Plants in Mined Caverns Study; the Downtown Seattle Transit Project, the Long Beach Naval Fuel Pier; and the Boston and San Francisco effluent outfall tunnels. He served on the committee that produced the recent NRC report *Improving Project Management in the Department of Energy* and as a reviewer of the NRC report *Assessing the Need for Independent Project Reviews in the Department of Energy*. He is the author of Parsons Brinckerhoff's *Project Management Manual* and also published various technical papers and articles. Mr. Gilbert is a member of the Board on Infrastructure and the Constructed Environment, Division on Engineering and the Physical Sciences, of the NRC and a variety of other organizations, including the American Society of Civil Engineers, the Project Management Institute, and the Society of American Military Engineers. He has won numerous awards in civil engineering and construction management, including being named a fellow of the American Society of Civil Engineers and receiving the Society's Rickey Medal and its Construction Management Award. He holds a B.S. in civil engineering and an M.S. in structural mechanics from the University of California, Berkeley, where he was also recently named Distinguished Engineering Alumnus.

Theodore C. Kennedy (National Academy of Engineering) is chairman and cofounder of BE&K, a privately held international design-build firm that provides engineering, construction, and maintenance for process-oriented industries and commercial real estate projects. Mr. Kennedy served as a member of the committee because of his experience and expertise with the design, construction, and cost estimation of complex construction and engineering projects. BE&K companies design and build for a variety of industries, including pulp and paper, chemical, oil and gas, steel, power, pharmaceutical, and food processing. BE&K is consistently listed as one of Fortune magazine's Top 100 Companies to Work For, and BE&K and its subsidiaries have won numerous awards for excellence, innovation, and programs that support its workers and communities. Mr. Kennedy is the chairman of the national board of directors of INROADS, Inc., and is a member of numerous other boards, including the A+ Education Foundation and the Community Foundation of Greater Birmingham. He is also a member of the Duke University School of Engineering Dean's Council and the former chairman of the Board of Visitors for the Duke University School of Engineering. He is a former president of Associated Builders & Contractors and a former chairman of the Construction Industry Institute. He has received numerous awards, including the Distinguished Alumnus Award from Duke University, the Walter A. Nashert Constructor Award, the President's Award from the National Association of Women in Construction, and the Contractor of the Year award from Associated Builders and Contractors. Mr. Kennedy has a B.S. in civil engineering from Duke University.

Michael A. Price is manager of education programs for the Project Management Institute (PMI), an international association of project management professionals that provides accreditation and training. He was appointed to this committee because of his experience and expertise in developing and evaluating project management training programs. Dr. Price is responsible for the development and implementation of operational plans for all PMI educational programs and initiatives, including accreditation of degrees in project management; selection and coordination of 150 public seminars annually; management of continuing education requirements and record keeping for 22,000 project management professionals; and identification of new educational products and programs to meet the learning needs of the global project management community. Previous to his present position, Dr. Price was director of professional practice for the American Institute of Architects (AIA) and director of programs for architecture and engineering with the Research Center for Continuing Professional and Higher Education at the University of Oklahoma. He is an active member of the AIA and has been a member of the Education System Audit Review Task Group and the site visitation team for the National Architectural Accreditation Board. Dr. Price has a B.S. in environmental design, a B.Arch., an M.Ed., and a Ph.D. from the University of Oklahoma.

B

Committee Fact Finding and Briefing Activities Through August 2001 and Project Data Request in March 2001

COMMITTEE ACTIVITIES

2000

July	Begin Phase III of oversight and review of DOE project management
September 19-20	Committee meeting 1. Overview of the status and direction of DOE project management by OECM, EM, DP, and SC, Washington, D.C.
	Briefings by Merna Hurd, senior advisor; Michael Telson, CFO; Clair Gill, director OECM; James Rispoli, deputy director OECM; Marvin Garcia, EM; Daniel Lehman, SC; Willie Clark, DP; and James Anderson, director SNS.
October 17-19	Observed OECM project management workshop and awards program (Al Burman, Lloyd Duscha, Brian Estes, Mike Price, and Mike Cohn), Arlington, Va.
November 15-17	Committee meeting 2. Response to follow-up questions from September presentations by OECM, EM, DP, and SC.
	Briefings by Clair Gill, director OECM; James Rispoli, deputy director OECM; Marvin Garcia, EM; James Carney, SC; and Willie Clark, DP.

November 17   Observe DP project management workshop (Don Brand), Washington, D.C.

2001

January 17   Release of interim letter report on project management improvements in DOE.

February 21-23   Committee meeting 3. Review the status of the project management initiative and of current projects in the Albuquerque Operations Office.

Briefings by Robert McMullan, OECM; Willie Clark, DP; and Marvin Garcia, EM.

Albuquerque Operations Office briefings by Albert Whitman, Technology and Site Programs; John Author, Environmental Operations and Services; Jack Tillman, Construction and Engineering; and George Rael, Environmental Restoration.

Site and project briefings for environmental projects by Beth Oms, Sandia National Laboratories (SNL); James Nunz, Kirkland Area Office; and Johnnie Guelker, Albuquerque Area Office.

Site and project briefings for DP projects by Pam McKever, SNL, Underground Reactor Facility; Wayne Evelo, KAO; David Post, LANL, Strategic Computing Complex; Mike Fulford, LAAO, Isotope Production; Lloyd Smith, LAAO, Accelerator Production of Tritium, Accelerator Applications; Sam Espinosa, APT Project Office; Linda Holland, Honeywell, Structural Upgrades and Strategic Stockpile Management Restructuring; Bob Schmidt, KCAO; and Robin Madison, BWXT Pantex.

Other briefings by Gregory Howell, Lean Construction Institute, and John Pearman, Energy Facilities Contractors' Group (EFCOG).

February 26-28   Presentation and discussions with DOE personnel, EM contractors, and EIR contractors (Ken Reinschmidt, Brian Estes, David Ford, and Mike Cohn) at Environmental Management 2001 conference, Tucson.

March 20   Meeting with Kevin Kolivar, Policy Advisor to Secretary Abraham (Ken Reinschmidt, Lloyd Duscha, Richard Little, and Mike Cohn).

March 21   Meeting and presentation with Energy Facilities Contrac-
                tors' Group (EFCOG) board (Ken Reinschmidt, Lloyd
                Duscha, and Mike Cohn).

March 26-28   Presentation and informal discussion at the Energy
                Monitor conference (Ken Reinschmidt, Brian Estes,
                Alan Burman, and Richard Little), Albuquerque.

April 3   Informal meeting with Richard Hopf on DOE contracting
                initiatives (Ken Reinschmidt, Lloyd Duscha, Allan
                Burman, Richard Little, and Mike Cohn).

April 4-6   Committee meeting 4. Review procedures for front-end
                planning by EM, DP, and SC and update on DOE
                project management issues in OECM, OMB, and
                Congress, Washington D.C.

                Briefings by Jeannie Wilson, U.S. House of Representa-
                tives Appropriations Committee; Laren Uher, Office of
                Federal Procurement Policy; and Walter Howes, DOE
                Office of Contract Reform and Privatization.

                Office of Environmental Management briefings by James
                Owendoff, Office of the Assistant Secretary; William
                Murphie, Office of Site Closure; Michael Weis, Office
                of Project Completion; Theresa Fryberger, Office of
                Science and Technology; Gene Schmidt, Planning and
                Budget; Pattie Bubar, Office of Integration and Dispo-
                sition; and Marvin Garcia, Office of Project Manage-
                ment.

                Office of Science briefings by Dan Lehman, Construction
                Management Support Division; Jim Carney, Construc-
                tion Management Support Division; Jeff Hoy, Materials
                and Sciences Engineering; Mike Riches, Office of
                Biological and Environmental Research; and Barry
                Sullivan, Laboratory Infrastructure Division.

                Defense Programs briefings by Willie Clark, Office of
                Project Management Support; Joel Leeman, Military
                Application and Stockpile Operations, Research, Devel-
                opment & Simulation; and Shah Jaghoory, Office of
                Facilities Management and ES&H Support.

                Office of Engineering and Construction Management
                briefings by Clair Gill, Dave Treacy, and Thad
                Kopnicki.

May 1   Meeting with Secretary Abraham, Kevin Kolivar, and Jim
                McSlarrow (Ken Reinschmidt, Lloyd Duscha, Richard
                Little, and Mike Cohn).

May 2-3	Presentation and participation in Office of Science project management workshop (Ken Reinschmidt, Lloyd Duscha, and Mike Cohn), Germantown, Md.
May 8-9	Observation of OECM roundtable review of O413.3, PPM manual, and PARS (Brian Estes, Lloyd Duscha, and Mike Cohn), Arlington, Va.
May 10-11	Informal review of project planning with DP20, DP10, and OS (Lloyd Duscha, Jeff Bostock, and Mike Cohn), Washington, D.C., and Germantown, Md.
May 15-17	Observation of SNS project review (Jeff Bostock), Oak Ridge, Tenn.
June 6-7	Presentation and participation at Defense Program project management workshop (Lloyd Duscha and Paul Gilbert), Las Vegas, Nev.
June 22	Meeting with Deputy Secretary Francis Blake (Ken Reinschmidt, Lloyd Duscha, Allan Burman, Richard Little, and Mike Cohn).
June 26	Meeting with David Swindel, chairman, EMAB Contracts and Management Committee (Lloyd Duscha and Mike Cohn).
July 11-13	Committee meeting 5, deliberation and report writing, Woods Hole, Mass.
July 18	Observed Project Management Career Development Task Force meeting (Lloyd Duscha), Washington, D.C.
August 3	Meeting with Bruce Carnes, CFO (Ken Reinschmidt, Lloyd Duscha, Allan Burman, Richard Little, and Mike Cohn).

PROJECT DATA REQUEST – MARCH 2001

1. All data packages and/or presentation packages which were provided for the CD-0, the CD-1, the CD-2, and the CD-3 for the following projects:

00-D-103 (LLNL) DP /TERASCALE SIMULATION
00-D-105 (LANL) DP /SCC
00-D-107 (SNL) DP/ JCEL
00-D-401 (SR) EM /SPENT FUEL TREATMENT & STORAGE TITLE I & II
01-E-300 (ORNL) OS /LAB FOR COMPARATIVE GENOMICS
01-D-124 (Y-12) DP /HEU STORAGE
01-D-126 (PX) DP /WEAPONS EVALUATION TEST LAB
01-D-407 (SR) MATERIALS DISP & EM /HEU BLEND DOWN
01-D-142 (VAR) EM /IMMOBILIZATION & ASSOC. PROCESSING

01-D-403 (RICH) EM /IMMOBILIZATION OF HIGH LEVEL WASTE
01-D-416 (RICH) EM /TANK WASTE

2. A tabulation of dollars expended by project/by fiscal year on project planning for the time prior to CD-0, the time between CD-0 and CD-1, and the time between CD-1 and CD-2.

Note: The committee assumes that most of the cost was expended by the contractors; however, if DOE had field office or headquarters cost, please include. Provide the critical decision dates. This tabulation should be provided for each of the above listed line items.

C

Phase II Report
Findings and Recommendations

DOE's portfolio of projects is large, complex, and sophisticated. Many projects are one of a kind, involving unique systems, processes, and technical challenges. Delivering projects of this magnitude that meet baseline costs and schedules is a constant challenge that requires excellent management. The findings and recommendations that follow provide guidelines for lifting DOE's project management to a level commensurate with other agencies and private industry. No single change will raise DOE's project management to the level required for such vital and expensive projects, because the problems are pervasive and cultural, and resolving them will require more than a quick fix. DOE must undertake a broad program of reform for the entire project management process.

This program of reform is set out in the recommendations, culminating in the recommendation that an office of project management be established to implement these reforms and drive cultural changes in DOE. To be effective, the proposed project management office must include the staff necessary to support the project managers and must provide consistent methods and systems for cost estimation, risk analysis, contracting, incentives, change control, progress reporting, and earned value management. The reform will require full and continuing support of the Secretary of Energy to ensure the support of program offices, field offices, and the entire DOE project management organization.

NOTE: Reproduced from *Improving Project Management in the Department of Energy*, National Research Council, 1999. Washington, D.C.: National Academy Press, pp. 3-9.

Policies, Procedures, Documentation, and Reporting

Finding. DOE does not have adequate policies and procedures for managing projects. No single authority is responsible for enforcing or ensuring that project management tools are used.

Finding. DOE has developed comprehensive practice guidelines for the design and construction phases of projects but has not developed comparable guidelines for the early conceptual and pre-conceptual phases, when the potential for substantial savings is high.

Finding. Many DOE projects do not have comprehensive project management plans to define project organization, lines of authority, and the responsibilities of all parties.

Finding. DOE does not effectively use value engineering to achieve project savings, even though federal agencies are required to do so.

Finding. DOE project documentation is not up to the standards of the private sector and other government agencies.

Finding. DOE does not have a consistent system for controlling changes in project baselines.

Finding. DOE does not effectively use available tools, such as earned value management, to track the progress of projects with respect to budget and schedule.

Finding. ISO 9000 provides a certification process by which an organization can measure itself against its stated goals, but DOE has not obtained certification. The certification process would help DOE remake the entrenched operating procedures and standards that have accumulated over the past 50 years.

Recommendation. As a part of its project management system, DOE should issue fundamental policies, procedures, models, tools, techniques, and standards; train project staff in their use; and require their use on DOE projects. DOE should develop and support the use of a comprehensive project management system that includes a requirement for a comprehensive project management plan document with a standard format that includes a statement of the project organization covering all participating parties and a description of the specific roles and responsibilities of each party.

Recommendation. DOE should update the project performance studies to document progress in these areas and extend the benchmarking baseline to include all major DOE construction projects. The study results should then be used to improve project procurement and management practices.

Recommendation. DOE should mandate a reporting system that provides the necessary data for each level of management to track and communicate the cost, schedule, and scope of a project.

Recommendation. DOE should establish a system for managing change that provides traceability and visibility for all baseline changes. Change control requirements should apply to the contractor, the field elements, and headquarters.

Recommendation. DOE should establish minimum requirements for a cost-effective earned-value performance measurement system that integrates information on the work scope (technical baseline), cost, and schedule of each project. These requirements should be included in the request for proposals.

Recommendation. DOE, as an organization, should obtain and maintain ISO 9000 certification for all of its project management activities. To accomplish this, DOE should name one office and one individual to be responsible for acquiring and maintaining ISO 9000 certification for the whole department and should require that consultants and contractors involved in the engineering, design, and construction of projects also be ISO 9000 certified.

Recommendation. DOE should establish an organization-wide value-engineering program to analyze the functions of systems, equipment, facilities, services, and supplies for determining and maintaining essential functions at the lowest life-cycle cost consistent with required levels of performance, reliability, availability, quality, and safety. Value engineering should be done early in most projects, and project managers should take the resulting recommendations under serious consideration.

Project Planning and Controls

Finding. DOE preconstruction planning is inadequate and ineffective, even though preconstruction planning is one of the most important factors in achieving project success.

Finding. DOE often sets project baselines too early, usually at the 2- to 3-percent design stage, sometimes even lower. (An agreement between Congress and DOE's chief financial officer for establishing baselines at the 20- to 30-percent design stage is scheduled to be implemented in fiscal year 2001.)

Finding. DOE often sets project contingencies too low because they are often based on the total estimated cost of a project rather than on the risk of performing the project.

Finding. DOE does not always use proven techniques for assessing risks of major projects in terms of costs, schedules, and scopes.

Recommendation. DOE should require that strategic plans, integrated project plans, integrated regulatory plans, and detailed project execution plans be completed prior to the establishment of project baselines. To ensure facility user and program involvement in the pre-construction planning process, DOE should require written commitments to project requirements from the ultimate users.

Recommendation. DOE should significantly increase the percentage of design completed prior to establishing baselines. Depending on the complexity of the project, the point at which project baselines are established should be between the completion of conceptual design and the completion of the preliminary design, which should fall between 10 and 30 percent of total design. The committee supports continuing efforts by Congress and the DOE to develop project baselines at a point of adequate definition beginning with fiscal year 2001.

Recommendation. Baseline validation should be assigned specifically to the project management office recommended in this report. The Military Construction Program of the U.S. Department of Defense, which requests planning and design funds for all projects in the preliminary design stage on the basis of total program size, is a potential model for DOE.

Recommendation. DOE should establish contingency levels for each project based on acceptable risk, degree of uncertainty, and confidence levels for meeting baseline requirements. The authority and responsibility for managing contingencies should be assigned to the project manager responsible for doing the work. In the process of evaluating potential projects, DOE should apply risk assessment and probabilistic estimating techniques, as required by the Office of Management and Budget.

Skills, Selection, and Training of Personnel

Finding. DOE's failure to develop project management skills in its personnel is a fundamental cause of poor project performance. DOE has shown little commitment to developing project management skills, as indicated by the lack of training opportunities and the absence of a project management career path. Successful organizations recognize that project management skills are an essential core competency that requires continuous training.

Recommendation. DOE should establish a department-wide training program for project managers. To ensure that this program is realistic, practical, and state of the art, DOE should enlist the assistance of an engineer/construction organization with a successful record of training project managers. DOE should establish criteria and standards for selecting and assigning project managers, including documentation of training, and should require that all project managers be trained and certified. DOE should also require that all contractors' project managers be experienced, trained, and qualified in project management appropriate to the project.

Project Reviews

Finding. Independent project reviews are essential tools for assessing the quality of project management and transferring lessons learned from project to project.

Finding. External independent reviews of 26 major projects are under way to assess their technical scope, costs, and schedules. The reviews so far have documented notable deficiencies in project performance verifying the committee's conclusion that DOE's project management has not improved and that its problems are ongoing. However, DOE has yet to formalize and institutionalize a process to ensure that the recommendations from these reviews are implemented.

Finding. Various DOE program offices are also developing the capability of conducting internal independent project reviews.

Recommendation. DOE should formalize and institutionalize procedures for continuing independent, nonadvocate reviews, as recommended in the Phase I report of the National Research Council to ensure that the findings and recommendations of those reviews are implemented. DOE should ensure that reviewers are truly independent and have no conflicts of interest.

Recommendation. All programs that have projects with total estimated costs of more than $20 million should conduct internal reviews, provided that the value of the reviews would be equal to or greater than the costs of conducting them. Deciding if an internal review is justified for a given project should be the joint responsibility of program management and the project management organization. The decision should be based on past experience with similar projects, the estimated cost of the project, and the uncertainty associated with the project. Internal reviews are expensive and take up the time of valuable people, so they should not be undertaken lightly. However, under the present circumstances, the committee believes that more internal reviews would be justified. The project management organization should manage these reviews for the director or assistant secretary of the cognizant program office. The results of these reviews should be taken by the program office to the Energy Secretary's Acquisition Advisory Board (ESAAB), and used as a basis for the decision whether to continue the project.

Acquisition and Contracting

Finding. Traditional DOE contracting mechanisms, such as cost-plus-award-fee and manage-and-operate (M&O) arrangements, are not always optimal for DOE's complex mission. These approaches are being replaced with more effective approaches based on objective performance incentives, but change has been slow.

Finding. DOE's long history of hiring contractors to manage and operate its sites on the basis of cost-plus-award-fee contracts has created a culture in which neither DOE nor its contractors is sufficiently accountable for cost and schedule performance.

Finding. DOE does not use effective performance-based incentives and does not have standard methods for measuring project performance.

Finding. DOE does not effectively match project requirements and contracting methods. Mismatching often results in cost and schedule overruns.

Finding. The numbers of bidders on major DOE contracts has been declining and in some cases have not elicited truly competitive bids. This may indicate that projects are not being appropriately defined and packaged and that the disincentives to bid often outweigh the incentives.

Recommendation. DOE should strengthen its commitment to contract reform focusing on the assessment and quantification of project uncertainties, the selection of the appropriate contract type and scope for each job, and increased use of performance-based incentive fees rather than award fees to meet defined project cost and schedule goals. A comprehensive risk analysis should be conducted before deciding whether to issue fixed-price contracts for work that involves a high level of uncertainty (such as new technology or incomplete characterization). Specific contract scopes and terms should be negotiated to define both DOE and contractor responsibilities to prevent cost overruns. Clear, written roles, authorities, and responsibilities should be established for DOE headquarters, field elements, contractors, and subcontractors for each contract. Guidelines should be provided for the appropriate times in the project for the selection of contractors.

Recommendation. DOE should develop written guidelines for structuring and administering performance-based contracts. The guidelines should address, but need not be limited to, the following topics: the development of the statement of work; the allocation of risks to whomever would be most effective at controlling the risks (either DOE or the contractor); the development of performance measures and incentives; the selection of the contracting mechanism; the selection of the contractor; the administration of the contract; and the implications of federal and DOE acquisition regulations. DOE should train its employees in the roles and responsibilities of a performance-based culture and then hold both employees and contractors accountable for meeting these requirements.

Recommendation. DOE should provide financial rewards for outstanding contractor performance to attract bids from the best contractors. A DOE-wide policy should be developed that provides fiscal rewards for contractors who meet or exceed schedule, cost, and scope performance targets. Contractor fees should be based on contractor performance.

Recommendation. DOE employees and contractor employees essential to projects should be trained in acquisition and contract reform. The training of source selection officials and members of source evaluation boards should be expedited; a minimum level of training should be a prerequisite.

Organizational Structure, Responsibility, and Accountability

Finding. DOE's organizational structure makes it much more difficult to carry out projects than in comparable private and public sector organizations. Successful corporations and agencies responsible for major projects arrange their organizations to provide focused and consistent management attention to projects.

Finding. Too many people in DOE act as if they were project managers for the same project, and too many organizations and individuals outside the official project organizations and lines of accountability can affect project performance.

Finding. Compliance with DOE's policy requiring the establishment of performance agreements and self-assessments from the field has been limited and slow.

Recommendation. To improve its project management performance, DOE should establish an office of project management on a level equal to or higher than the level of the offices of assistant secretaries. Department-wide project management functions should be assigned to the project management office, and the director of this office should have the authority and the resources to set and enforce reporting requirements for all projects. Other responsibilities, such as property and asset management, should be assigned to existing DOE headquarters offices. To be successful, the office of project management must have the full and continuing support of the secretary, the under secretary, the deputy secretary, and of all of the program offices and field offices as a top-down management initiative.

D

Letter Report of January 2000

The National Research Council (NRC) Committee on Oversight and Assessment of U.S. Department of Energy (DOE) Project Management has completed its initial assessment of DOE's progress in implementing the recommendations from the 1999 NRC report, *Improving Project Management in the Department of Energy* (the Phase II report), and related actions. The committee's assessment is based on briefings by DOE staff and others involved with improving project management, a review of documents provided by DOE, and other relevant materials. The review and assessment were directed by the 106th Congressional Committee of Conference on Energy and Water Development (House Report 106-336).

This letter report is submitted pursuant to an agreement between DOE and NRC for a report six months after initiation of the study. It transmits the committee's assessment of DOE's progress and recommends additional actions to further improve DOE project management capabilities. The letter consists of an overall summary, observations, findings, and recommendations relating to the general categories of recommendations in the Phase II report.

The committee intends to seek further input from DOE headquarters, field offices, and projects, as well as from current, former, and potential DOE contractors, in subsequent efforts to determine how well project management reforms are working and what additional steps may be necessary for DOE to achieve

NOTE: Reproduced from "Improved Project Management in the Department of Energy," letter from BICE Committee for Oversight and Assessment of U.S. Department of Energy Project Management Chair Kenneth F. Reinschmidt to Secretary of Energy Spencer Abraham (January 17, 2000).

excellence in project management. The committee appreciates the cooperation and support of the Office of Engineering and Construction Management (OECM), the project management support offices (PMSOs), and the other elements of DOE.

SUMMARY

DOE has undertaken a number of initiatives to improve project management since the 1999 NRC report, *Improving Project Management in the Department of Energy* (the Phase II report), was published. In 1999, DOE established OECM and the PMSOs in three program secretarial offices (PSOs): the Office of Science, the Office of Defense Programs, and the Office of Environmental Management. The release of DOE Order O413.3, " Program and Project Management for the Acquisition of Capital Assets," and the DOE-wide Program and Project Management 2000 Workshop, both in October 2000, were also notable steps in the right direction and indicative of greater interest and involvement on the part of the deputy secretary and the chief financial officer (CFO) in project management.

As stated in the Phase II report, effective and accountable project management should be a continuing priority for DOE and its leaders at all levels. Through actions taken to date, DOE has begun to address some of the core issues. However, a number of issues have not been resolved. The most important unresolved issues are: (1) the definition of the authority and scope of OECM; (2) the provision of adequate financial and staff resources to improve project management; (3) the development and implementation of contract performance-measurement systems; (4) the design and implementation of an information-management system that can track contracts and contractor performance and feed information back into key decisions; and (5) continued emphasis on close cooperation and trust within DOE and with its contractors that will be fundamental to the long-term effectiveness of project-management reforms.

Although the committee considers the organizational changes made so far as generally positive, they are only beginnings. In the 18 months since the Phase II report was published, DOE could not possibly have implemented all of the necessary project-management reforms or achieved a high level of excellence. Much more time and attention will be necessary to achieve the goals set out in the Phase II report, and the committee recognizes that, until reforms have taken effect throughout the organization, project-management failures can be anticipated. As stated in the Phase II report, there is no "quick fix" for DOE's problems. Improving project management in DOE will require changes in organizational structures, documents, policies, and procedures, as well as substantial changes in the culture of the department. In order to be effective, these changes must be embraced at all levels of the organization, especially in field and project offices.

Based on information provided by DOE, the committee believes that OECM and the PMSOs do not have adequate resources to perform their many functions effectively, particularly in light of the high costs, complexity, and urgency of DOE projects and the great need for improved project management. To ensure that the necessary changes and improvements are made, the committee strongly recommends that the authority of OECM and the PMSOs be strengthened and that the resources and personnel available to them be increased to support their responsibilities. By strengthening the roles of OECM and the PMSOs, DOE can establish a strong in-house center of excellence that will ensure the implementation of improved project-management procedures.

To strengthen and affirm DOE's commitment to reforming its project management, the committee reiterates the recommendation in the Phase II report that OECM be the unifying organization for project management throughout the department. OECM should be at the level of assistant secretary and report directly to the deputy secretary of energy. This would promote consistency and commitment throughout the department and encourage a culture of excellence in project management.

ORGANIZATIONAL STRUCTURE, RESPONSIBILITY, AND ACCOUNTABILITY

Subsequent to the publication of the Phase II report, DOE established OECM within the office of the department's CFO and the PMSOs in the three major PSOs. Their responsibilities were defined in DOE Order O413.3. The committee believes that these positive steps can lead to significant improvements in project performance. Nevertheless, this organizational structure differs significantly from the one recommended in the Phase II report, which endorsed a strong, central, project-management office reporting directly to the deputy secretary. Although the PMSOs are positive additions to the new project-management structure, the committee believes that OECM would have a greater positive impact if it were elevated to the level of assistant secretary and reported directly to the deputy secretary; this would establish a peer relationship among OECM and the PSOs while maintaining consistent professional leadership for the office. The committee also believes that DOE project management and OECM would be more effective if the following OECM responsibilities were included in Order O413.3:

- Specify project-reporting requirements.
- Define and implement a DOE project-management information reporting system.
- Review all projects and validate that they are in compliance with the DOE project policies and procedures, and initiate actions to correct noncompliant practices.
- Review and validate proposed variations in project-management procedures to ensure continued compliance with the established objectives.
- Initiate and maintain a database of project-management experiences for the department.

- Advise the deputy secretary of all matters related to projects and project management.

SKILLS, SELECTION, AND TRAINING OF PERSONNEL

Competent project-management professionals are essential to successful projects. The committee affirms the recommendation in the Phase II report that the department institute an effective career-development program to facilitate the recruitment, development, and retention of competent, professional project managers. OECM has developed a plan to create a department-wide career-development program and has received funds to carry out the planning phase. The committee does not have sufficient information to assess progress in training and professional development; however, an effective, widely implemented career-development program will require sufficient resources and support for full implementation.

The committee recommends that OECM ensure that the career-development program provides DOE personnel with access to a variety of learning resources and training methods and that the curriculum addresses competency in team building, DOE policies, and general project-management tools and techniques. DOE should foster a climate of learning and cultural change by supporting project-management personnel in obtaining professional certification and participating in professional activities. DOE should also encourage its contractors to support similar career-development efforts.

The implementation of an effective, department-wide, career-development program will be critical to improving DOE's project management. Therefore, the committee will continue to monitor the department's efforts in this area closely.

POLICIES, PROCEDURES, DOCUMENTATION, AND REPORTING

Policies and Procedures

The Phase II report recommended that DOE develop more effective project-management policies, procedures, models, tools, techniques, and standards; train staff in their use; and require their application for all DOE projects. The report also recommended that DOE develop a project-management system that includes a requirement for a standard project-management plan, including a statement of the project organization covering all participating parties and a description of the specific roles and responsibilities of each party.

To date, the efforts of the deputy secretary, OECM, the PSOs, and the PMSOs have unquestionably raised awareness of the importance of good project management. Briefings by representatives of the PMSOs on project-management procedures they have established reflect good coordination with OECM. If these activities are continued and extended, they could become the foundation of a coherent project-management approach for the entire department.

DOE Order O413.3 and drafts of *Program and Project Management Manual* (PPMM) and *Program and Project Management Practices* (PPMP) are evidence that a start has been made on improving project-management policies, procedures, models, tools, techniques, and standards. However, much remains to be done. Although the committee has not yet completed a comprehensive review of the PPMM and PPMP, a few general observations can be made at this time. (A more detailed review and assessment may be included in a future report.)

The PPMM and the PPMP are greatly improved over the previous DOE guidance documents, and the committee congratulates OECM and other contributors on their efforts. OECM has stated that they intend to revise and issue the documents as directives within the next year. If the documents are revised appropriately they could form a central framework for DOE's project-management capability, as recommended in the Phase II report. However, these documents should focus on defining how DOE does business, as opposed to general project-management methodologies, which should be incorporated by reference to texts and handbooks in the field. The sections on risk analysis and contingency in particular should be rewritten to reflect how DOE wants these procedures to be carried out and to promote a consistent approach throughout the department.

The effectiveness of policies, procedures, and models is determined by how consistently they are understood and supported by the individuals who carry them out. The committee found some indications that the PPMM and PPMP have been accepted throughout the organization. However, it is not clear who is responsible for verifying implementation of the policies and procedures. Project-management documentation should clearly define DOE's systems and processes, and expectations of senior management for project performance, as well as organizational and individual incentives for managers at all levels to pursue effective, accountable project management. The documents should clearly identify a staff position responsible for verifying policy implementation and quality assurance. The committee believes that this oversight would be an appropriate function for OECM.

Neither the PPMM nor the PPMP defines the terms *program* and *project* as they are used by DOE. In fact, the terms *program* and *project* are used interchangeably. Although this may not create an immediate problem, the application of the policy documents may require that the difference between *programs* and *projects* be clearly understood. A *project* is usually a specific set of tasks, with a beginning, a middle, and an end. A *project* also has a well defined scope, cost, and schedule. Thus, a *project* is likely to be a controllable effort, the progress and performance of which can be assessed using standardized methods. A *program* is usually a group of projects. The complex scope and extended duration of a *program* can be made manageable by subdividing the whole into definable, understandable, controllable units or *projects*. A *program*, however, is more than the sum of its *projects* because each *program* must respond to the specific mission and integrate *projects* into a working whole. For example, the risks and contingencies for a *program* are not simply the sum of the risks and contingencies for the projects.

Reporting

The Phase II report recommended that DOE develop and implement a comprehensive project reporting system. The committee notes that the PMSOs are reporting some project data and that OECM has established a general target for reporting practices. However, current reporting requirements, tools, and practices are still incomplete and inconsistent among projects and programs. The lack of standard tools and procedures has prevented the aggregation of project data that could be used to evaluate project performance at the program and departmental levels. A consistent, reliable project-reporting system will be critical to achieving excellence in project management. Therefore, OECM should develop specific, precise requirements for integrated project and program reporting. In addition, OECM should provide training and support services to the PMSOs for reporting and collecting project data. Insufficient progress has been made in the development of an effective reporting system for the committee to offer a more detailed assessment, but project reporting should be given a high priority.

Change Control

The Phase II report recommended that DOE develop and implement a comprehensive change-management system. To date, DOE has defined a target process for change management in DOE Order O413.3, and the committee is eagerly awaiting the implementation of the proposed process and looking forward to an opportunity to assess its effects on projects. An effective change-management system is critical in the prevailing DOE cost-plus environment, as well as for fixed-price and lump-sum projects. Change-management processes, including reviews by the Energy Systems Acquisition Advisory Board (ESAAB) and the change-control boards (CCBs), will be evaluated when sufficient data on experiences with change-control practices in actual projects have been collected.

Earned Value Management

The Phase II report recommended that the DOE use an earned value management system (EVMS)[1] to track project performance. EVMS has been identified as a primary project-management procedure in DOE Order O 413.3. Some individual projects have already reported earned value data in their quarterly reports,

---

[1]Earned value management is a method of making an objective assessment of performance by relating the actual cost of work performed (earned) to budgeted costs.

and DOE has awarded honors to three projects that have used EVMS. The committee acknowledges progress in this area and reiterates the importance of an earned value approach for project management. A consistent, earned value management approach would provide DOE project managers, program managers, and senior managers with an objective means of evaluating the status of projects, predicting future progress, and responding effectively to actual project conditions. Significantly more support from senior management including training, technical resources, and encouragement, will be necessary for EVMS to be implemented and used by DOE managers at all levels. The committee encourages DOE to use EVMS to predict project-performance outcomes and to manage projects proactively, as well as to report project status accurately.

ISO 9000 Certification

The Phase II report recommended that DOE obtain ISO 9000[2] certification, but no preparations for ISO 9000 certification have been initiated to date. However, OECM indicates that certification will be sought when measurable improvements have been made to DOE project-management processes and structures. Although DOE has not yet begun the formal process of obtaining ISO certification, DOE Order O413.3, the PPMM, and PPMP are appropriate beginnings for an ISO 9000 process. The committee recognizes that ISO 9000 certification by itself will not improve DOE's project management. However, the process involved in preparing for and seeking ISO 9000 certification will have direct benefits. For example, DOE will be required to purge the outdated and inconsistent policies, procedures, and regulations that have accumulated over the years and to focus on the essential elements of successful project management. The committee will continue to assess DOE's implementation plan for ISO 9000 certification.

Value Engineering

Although OECM has described value engineering (VE)[3] as a desirable practice, according to the DOE Inspector General,[4] it has not been widely or consistently used. DOE Order O413.3 lists OMB Circular A-131, "Value Engineering," as a reference and states that DOE is committed to using VE. The Contractor Requirements Document, Attachment 1 to DOE O413.3, states that a VE process

---

[2]ISO 9000 is a quality-performance standard established by the International Organization for Standardization (ISO). ISO 9000 has been widely embraced by private-sector and government organizations worldwide.

[3]Value engineering is an organized effort to analyze projects to achieve essential functions at the lowest life-cycle costs consistent with required performance.

[4]Audit Report HQ-B 98-01, DOE's Value Engineering Program.

must be used, and VE is mentioned in the draft PPMM and PPMP. Because DOE Order O413.3 was issued only recently, compliance cannot yet be assessed. However, the committee believes that DOE Order O413.3 does not define a VE process and does not define a process for verifying the effective use of VE.

Although DOE Order O413.3 is a positive step toward the implementation of VE, it can not, by itself, effectively make VE an integral part of DOE project management. DOE project managers have to be trained in the use and interpretation of VE, and a certified VE specialist should be appointed to oversee and promote its application.

PROJECT PLANNING AND CONTROLS

The committee is encouraged by an agreement between Congress and DOE to establish project baselines after 20 to 30 percent of a project design has been completed and the creation of a funding mechanism for project planning, engineering, and design (PED). As noted in the Phase II report, adequate PED funding, preproject planning, and project controls are all critical to successful projects. The committee encourages DOE to continue implementing procedures to establish project baselines at an appropriate level of design completion and to implement other measures to improve the accuracy and reliability of cost and schedule estimates.

DOE has developed a fairly detailed project-planning process as part of its capital budget cycle, which should promote effective planning of projects. Other tools, such as checklists, communications software and methods, planning reviews, third-party audits, economic modeling, setting of measurable objectives, and team building, can also help. The committee believes that objective evaluations of new technology, and information-flow and work-flow design should be made during the project-planning phase.

OECM, in conjunction with the PMSOs, has begun to develop some of these project-planning initiatives. OECM has already documented some planning procedures and should revise and expand the descriptions in the PPMM and PPMP. These documents should also reference appropriate, up-to-date sources of project-planning methodologies. The PMSOs should provide supporting policies and procedures tailored to the specific projects and needs of their programs, as well as oversight to ensure quality; OECM should validate project plans prior to critical decision points. Procedures should be established to ensure that projects are not unnecessarily delayed by poor plans and that time pressures do not lead to projects being approved without adequate planning. All members of the project team should review project plans and provide written commitments and concurrence on the project scope, cost, and schedule.

The primary responsibility for the planning phase of project development lies with DOE personnel. Contractor assistance should be sought as needed. Even when a planning process is in place, it is the responsibility of DOE management to ensure that every project is planned effectively. This monitoring could be accomplished through process audits, performance benchmarking, and direct observation and interaction with project teams. Project-team members should be held accountable for project planning and subsequent performance, and projects in trouble must be identified early—not in the late execution phase. Senior management can ensure that effective project planning is being conducted in the following ways:

- asking questions at project review meetings
- providing resources to support process training and implementation
- ensuring strategic flexibility (including cost and schedule contingencies)
- maintaining discipline in sticking to the plan
- benchmarking results

The committee recognizes that it will take time before consistent preproject planning can be integrated into project management throughout the organization. Preproject planning will require both procedural and cultural changes. However, DOE management should make it known that effective preproject planning will be required for all projects, without exception. Training or proof of proficiency in preproject planning should be required of all project-team members prior to the start of new projects.

PROJECT REVIEWS

External Independent Reviews

Language in the *Energy and Water Development Appropriations Bill, 2001*, indicates that Congress relies heavily on external independent reviews (EIRs) for objective project evaluations. As a result, the number of EIRs has increased perceptibly in the past two years. Although EIRs are, overall, useful to DOE and to DOE projects, EIRs that provide only general information are of limited value. Some reviews have even provided inaccurate and misleading conclusions[3], raising questions about the competence and independence of the reviewers. Some deficiencies can probably be attributed to inadequate definitions of the scope of the reviews and a lack of understanding of the fundamental goals of the review.

The committee believes that the EIR program is important but that it requires some modification. In view of the emphasis on EIRs by Congress, DOE should ensure that this program is effective. OFCM has reported taking some steps to establish procedures, goals, and expected results for EIRs. However, the docu-

---

[3]National Ignition Facility, Management and Oversight Failures Caused Cost Overruns and Schedule Delays. GAO/RCED-00-271.

mented policies and procedures have not been reviewed or evaluated by the committee.

OECM should develop quality standards for EIRs and monitor projects to ensure that the reviews are conducted properly. All concerns raised during project reviews should be well documented and satisfactorily addressed. The committee reiterates the Phase II recommendation that OECM ensure that reviewers are truly independent and have no conflicts of interest. DOE should formally evaluate reviewers and use the evaluations as references in the selection of future independent reviewers.

Internal Reviews

Congressional requirements also mandate that all line-item projects be reviewed before any new money is spent. The congressional requirements go even further than the recommendations for internal reviews in the Phase II report. DOE had a history of conducting internal reviews even prior to the Phase II report. The most formalized and intensive internal review process has been developed by the Office of Science, which recently released a draft *Independent Review Handbook* documenting its approach. The committee has not reviewed this document in sufficient detail to evaluate it at this time.

Although Congress has promoted internal independent reviews, it is not clear to what extent current DOE procedures have addressed congressional concerns. Although internal reviews are not currently managed centrally, as was recommended in the Phase II report, OECM has been involved in a support role. The congressional emphasis on internal reviews and their potential for ensuring project success warrant the development of procedures and guidelines for all of the PSOs, not just the Office of Science. The specific missions of the program offices may require different internal review procedures, but the fundamental goals and objectives of internal reviews should be identical. In the absence of department-wide control of the internal review process, the PMSOs should formalize a coordinated process to facilitate central oversight and the transfer of lessons learned among programs. OECM should also evaluate the effectiveness and economic justification for internal reviews of small projects.

ACQUISITION AND CONTRACTING

OECM has taken a number of actions that could improve DOE acquisition and contracting processes. DOE Order O 413.3 enumerates the steps to be followed in preconcept planning, risk analysis, and the overall acquisition process. Many sections of the draft PPMM also address these issues. Although documents can provide useful guidance, success will be determined by how well these procedures are followed and the willingness of all participants in the contracting process to develop appropriate contracting types, terms, and conditions.

DOE has stated that the major vehicle for improving acquisition and contracting is the integrated project team (IPT), which is made up of key staff from the contracting office, the program and project offices, and OECM. Although OECM does not have direct project responsibilities, it should be an active participant in the IPT, which should address the risks and uncertainties that have plagued previous DOE contracting efforts.

Because the IPT bears much of the front-end responsibility for contracting new projects, team members must be well versed in using performance-based contracting (PBC), which is designed to focus all participants on the critical issues, such as risk, uncertainty, and accountability. IPT members should be skilled in using PBC techniques with appropriate performance standards and incentives and an effective system for assessing contractor progress. OECM should designate a senior staff member to become an authority on PBC. The committee notes that OECM has not yet focused on this area and recommends that DOE develop a formal process for identifying the need for training in PBC, as well as staff to be trained. The objective should be to create an acquisition workforce of both program and contracting personnel capable of using innovative contracting methods. The committee will continue to assess DOE acquisition and contracting reforms and the application of these reforms to projects and programs including management-and-integration (M&I) and management-and-operations (M&O) contractors.

CONCLUSIONS

In general, the committee commends DOE for taking positive steps toward improving its project-management capabilities. Much remains to be done, however, and the committee will continue to review and assess DOE's progress. The committee appreciates this opportunity to be of service to DOE and looks forward to assisting the department in its continuing efforts to improve project performance.

COMMITTEE FOR OVERSIGHT AND ASSESSMENT OF U.S. DEPARTMENT OF ENERGY PROJECT MANAGEMENT

KENNETH F. REINSCHMIDT, *chair*, Stone and Webster, Inc. (retired), Littleton, Massachusetts

DON JEFFREY BOSTOCK, Lockheed Martin Energy Systems (retired), Oak Ridge, Tennessee

DONALD A. BRAND, Pacific Gas and Electric Company (retired), Novato, California

ALLAN V. BURMAN, Jefferson Solutions, Washington, D.C.

LLOYD A. DUSCHA, U.S. Army Corps of Engineers (retired), Reston, Virginia

G. BRIAN ESTES, Consulting Engineer, Williamsburg, Virginia
DAVID N. FORD, Texas A&M University, College Station, Texas
G. EDWARD GIBSON, University of Texas, Austin
PAUL H. GILBERT, Parsons Brinckerhoff International, Inc., Seattle, Washington
PATRICIA W. INGRAHAM, Syracuse University, Syracuse, New York
THEODORE C. KENNEDY, BE&K, Inc., Birmingham, Alabama
MICHAEL A. PRICE, Project Management Institute, Newtown Square, Pennsylvania

E

Statistical Process Control with EVMS Data

The standard EVMS reporting quantities, cumulative through reporting period t (typically in months), are defined as follows:

$BCWS(t)$ = cumulative Budgeted Cost of Work Scheduled through reporting period t

$BCWP(t)$ = cumulative Budgeted Cost of Work Performed through reporting period t

$ACWP(t)$ = cumulative Actual Cost of Work Performed through reporting period t

$SPI(t) = BCWP(t)/BCWS(t)$ = cumulative Schedule Performance Index through reporting period t

$CPI(t) = BCWP(t)/ACWP(t)$ = cumulative Cost Performance Index through reporting period t

To apply control charting methods,[1] it is necessary to consider the earned value quantities in a single reporting period t, written as:

$bcws(t)$ = incremental Budgeted Cost of Work Scheduled in reporting period t

$bcwp(t)$ = incremental Budgeted Cost of Work Performed in reporting period t

$acwp(t)$ = incremental Actual Cost of Work Performed in reporting period t

---

[1]See, for example, Forrest W. Breyfogle III. 1999. Implementing Six Sigma: Smarter Solutions Using Statistical Methods. New York, N.Y.: John Wiley & Sons, pp. 519-525.

spi(t) = bcwp(t)/bcws(t) = incremental Schedule Performance Index
cpi(t) = bcwp(t)/acwp(t) = incremental Cost Performance Index

The cumulative and incremental definitions are linked by the following:

bcws(t) = BCWS(t) − BCWS(t − 1) or BCWS(t) = BCWS(t − 1) + bcws(t)
bcwp(t) = BCWP(t) − BCWP(t − 1) or BCWP(t) = BCWP(t − 1) + bcwp(t)
acwp(t) = ACWP(t) − ACWP(t − 1) or ACWP(t) = ACWP(t − 1) + acwp(t)

That is, acwp(t) is the actual cost of work performed in the time period t, whereas ACWP(t) is the cumulative actual cost of the work performed from the beginning of the project through time period t. The time period is the reporting period, usually a month.

Due to random fluctuations in project conditions, the dimensionless indices spi(t) and cpi(t) will vary. These statistical variations are typically assumed, for convenience, to be drawn from normal distributions, with some defined means and standard deviations. If the project is in a state of statistical control, these means and variances will be stable. The mean values of spi(t) and cpi(t) should be 1.0 (greater than 1.0 is better; less is worse) and the variances of both should be acceptably small. Zero variances, which would indicate exceptional quality of project planning and control, are unlikely to occur. The objective of the analysis is to determine whether a project has gone out of statistical control, which means that either the mean of spi(t) or cpi(t) is changing or the variance is changing, or both. And if a project is going out of control, this may mean that the project will go over schedule or over budget in the future.

To evaluate whether a change is occurring in the mean (the measure of central tendency) of spi(t), one should first establish the average value and the standard deviation based on historical data from projects that are considered to have been under control and good performers. Then, upper and lower natural process limits (UNPL and LNPL), which are conventionally three standard deviations above and below the mean, are derived. This usage is the source for the well-known six-sigma limits in statistical process control charting (one sigma is the standard deviation). Then, the probability that the measured spi(t) will be below the three-sigma LNPL (based on the normal distribution) owing to statistical fluctuations alone is 0.0013, and the probability that spi(t) will be above the UNPL is also 0.0013. That is, if the measured spi(t) is below the LNPL, this indicates that the process may be going out of control, as the probability that this value would occur with the process in control is only about 1/1,000. More specifically, if management were to follow up on every value of spi(t) below the LNPL to investigate a possible adverse change in the process, management would be wrong only once in a thousand times.

As a simple indicator of dispersion or variability, control charting methods often use the period-to-period range, which is the absolute magnitude of the difference between the current period value and that in the previous period—for example:

spirange(t) = | spi(t) – spi(t – 1) | for the schedule performance index
cpirange(t) = | cpi(t) – cpi(t – 1) | for the cost performance index

The mean and the variance for the range can be determined by statistical methods,[2] and the upper control limit and the lower control limit for the range established as the mean plus or minus three standard deviations, as before. Then, if spi range(t) is observed above the upper control limit, there is only one chance in a thousand that this would occur through random fluctuation if the process is in control, and it is more likely due to a change in the variability of the process.

Note that the mean of the process, E[spi(t)], could be changing with no change in the variance, or vice versa. Also, some changes are beneficial: a decrease in E[spi(t)] may indicate that the project will fall behind schedule, but an increase in E[spi(t)] is favorable. Similarly, a reduction in the standard deviation of spi(t) means that the project is under tighter control, whereas an increase in the standard deviation may be indicative of future problems.

Figure E-1 shows the incremental (monthly) values for spi(t) for the Tritium Extraction Facility, along with the mean and the natural process limits. These are the same EVMS data as shown in Figure 5-1 but on an incremental, monthly basis. In this case, the process limits were computed using this project only, so the plot merely indicates that the project is staying within itself. As shown in the earned value plot (Figure 5-1), this project is close to schedule and budget at the last shown reporting date, so presumably is not in trouble, and the plot of monthly spi(t) corroborates this. A project in trouble would show values outside the process limits or other indications of changing conditions.

The same information can be shown on a cumulative or running SPI(t) plot, if preferred, as in Figure E-2. Here, the six-sigma band between the upper and lower process limits narrows with time, to compensate for the inertial effect of past history—that is, as t increases, a major change in spi(t) results in a smaller change in the cumulative SPI(t). Figure E-2 shows that although SPI(t) starts out below 1.0, it is still in the six-sigma band, so both plots give the same indication: the fluctuations (typically called "variances," but here the term "variance" is always used in the statistical sense, to mean "square of the standard deviation") are within the natural process limits, so no management attention is required.

Of course, the six-sigma process limits are simply points on the normal probability distribution and by themselves say nothing about quality. To use the measured data for quality control, one must determine the upper and lower specification limits (USL and LSL), the band of acceptable performance—that is, the

---

[2]Ibid., p. 523.

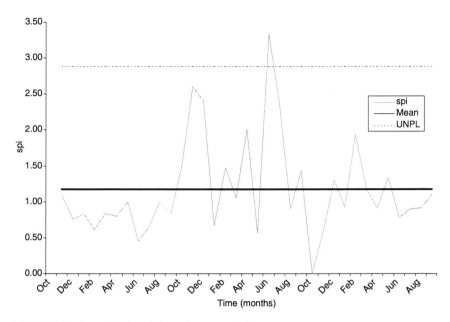

FIGURE E-1 Monthly Schedule Performance Index.

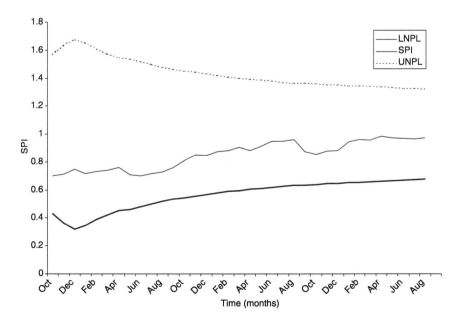

FIGURE E-2 Cumulative Schedule Performance Index.

band in which management specifies that the values should lie. Then, if LSL < LNPL < UNPL < USL, the process lies within the specification requirements, quality is acceptable (this is six-sigma quality), and management should be happy. If LNPL < LSL < USL < UNPL, the process lies outside the specification requirements, quality is unacceptable, and management should be taking action.

A comparable metric is the capability index, Cp, which may be defined as Cp = (USL – LSL)/(six-sigma). If Cp < 1, the process is not very capable; that is, it cannot produce acceptable quality.

Of course, one does not have to place the limits at three standard deviations above and below the mean. Generally, six-sigma quality is regarded as excellent quality. Not every process is excellent, and this includes the DOE project management process. For example, consider the Tritium Extraction Facility plots in Figures E-1 and E-2. In these, the process limits were computed from the variations in this single project. No attempt is made here to set the specification limits, but the values are so highly variable (i.e., the standard deviation is so high) that the six-sigma process limits are extremely broad. It is very likely that acceptable specification limits would lie well inside this band, meaning that the variability in this project may be consistent but is too high to be considered good quality.

Figures E-1 and E-2 relate to the schedule performance index spi(t) and SPI(t), but identical arguments apply to the cost performance index cpi(t) and CPI(t).

In addition, there are many other variables that could provide management with indicators of potential project problems. One example would be change orders across all projects in one PSO; the Phase II report expressed the opinion that changes on DOE projects were out of control, and charting changes over time would indicate whether DOE is improving.

From the perspective of OECM, there are a sufficient number of projects in DOE to provide the statistical basis for applying six-sigma control charting as a filter, to identify projects that may be trending out of control. In fact, each of the three major Program Offices has enough projects to do this separately. The value of the statistical charting method is that it would put the evaluation of projects throughout the complex on a consistent, scientific, objective basis. There is, of course, vastly more about interpretation of control charts than is discussed here (see Breyfogle, for example, or any good textbook on statistical process control); the necessary computations are not presented here, but they are easily performed automatically on a computer.

It should be reemphasized that, for any charting and analysis to occur, the values must be measured and reported. That is, if there is ever to be any improvement, the project reporting system must be designed to support the needs of analysis.

Statistical studies have shown that the use of the cumulative CPI(t) provides an efficient estimator for the estimated cost at completion (ECAC). That is, from

$$CPI(t) = BCWP(t) / ACWP(t)$$

one derives

$$ECAC = BAC / CPI(t)$$

where BAC = budget at completion. There is no corresponding equation for the estimated date at completion (EDAC) in general use. However, there are methods for estimating EDAC from $BCWS(t)$ and $BCWP(t)$—for example, linear and nonlinear regression—that could be evaluated for use in PARS. Even regression is easily performed by computer. Obviously, any forecast of EDAC and ECAC depends on accurate reporting without rebaselining. If a project is continually rebaselined such that $CPI(t) = 1$, the above equation will always give ECAC = BAC.

F

Acronyms and Abbreviations

ACWP	actual cost of work performed
AE	acquisition executive
AO	Albuquerque Operations Office
AQL	acceptable quality level
ARM	acquisition risk management
BAC	budget at completion
BCWP	budgeted cost of work performed
BCWS	budgeted cost of work scheduled
BICE	Board on Infrastructure and the Constructed Environment
CAP	corrective action plan
CD-0	critical decision 0, approval of mission need
CD-1	critical decision 1, approval of preliminary baseline range
CD-2	critical decision 2, approval of performance baseline range.
CD-3	critical decision 3, approval of start of construction
CD-4	critical decision 4, approval of start of operation or project closeout
CFO	chief financial officer
CII	Construction Industry Institute
CPI	cost performance index
CUI	contingency utilization index
DMAIC	define, measure, analyze, improve, control (six-sigma process)
DOE	U.S. Department of Energy

DP   National Nuclear Security Agency Office of Defense Programs

ECAC   estimated cost at completion
EDAC   estimated date at completion
EIR   external independent review
EM   Office of Environmental Management
ESAAB   Energy Systems Acquisition Advisory Board
EVMS   earned value management system

FAR   Federal Acquisition Regulations
FPM   federal project manager

GAO   U.S. General Accounting Office
GPRA   Government Performance and Results Act of 1993

ICE   independent cost estimate
ICR   internal cost review
IPR   internal project review
IPT   integrated project team
ISO   International Organization for Standardization

LANL   Los Alamos National Laboratory
LLNL   Lawrence Livermore National Laboratory
LNPL   lower natural process limits
LSL   lower specification limit

M&I   management and integration
M&O   management and operations

NNSA   National Nuclear Security Agency
NRC   National Research Council

OECM   Office of Engineering and Construction Management
OFPP   Office of Federal Procurement Policy
OMB   Office of Management and Budget, Executive Office of the
            President

PARS   Project Analysis and Reporting System
PBC   performance-based contracting
PDRI   project definition rating index
PED   project engineering and design (funding)
PEP   project execution plan
PMCDP   Project Management Career Development Program

PMP   *Project Management Practices*
PMSO   project management support office
PPM   *Program and Project Management* manual
PRMS   project review management system
PSO   program secretarial office

RACE   risk-adjusted cost estimate

SC   Office of Science
SLAC   Stanford Linear Accelerator Center
SNL   Sandia National Laboratories
SOW   statement of work
SPC   statistical process control
SPI   schedule performance index

TEC   total estimated cost
TEF   Tritium Extraction Facility
TPC   total project cost

UNPL   upper natural process limit
USL   upper specification limit

VE   value engineering